Redesigning the American

Lawn

Redesigning the American

Lawn

A Search for Environmental Harmony

SECOND EDITION

F. Herbert Bormann

Diana Balmori

Gordon T. Geballe

LISA VERNEGAARD & SALLY ATKINS, *Editors-Researchers*

Yale University Press New Haven and London

Published with assistance from the foundation established in memory of
Amasa Stone Mather of the Class of 1907, Yale College.

Set in Monotype Garamond by Duke & Company, Devon, Pa.
Printed in the United States of America by Worzalla, New York, N.Y.

Library of Congress Cataloging-in-Publication Data
Bormann, F. Herbert, 1922–
    Redesigning the American lawn : a search for environmental harmony /
F. Herbert Bormann, Diana Balmori, Gordon T. Geballe.—2nd ed.
        p.   cm.
Includes bibliographical references (p.   ).
    ISBN 0-300-08694-6 (pbk. : alk. paper)
1. Lawns—United States. 2. Lawn ecology—United States. I. Balmori, Diana.
II. Geballe, Gordon T., 1947–   III. Title.
    SB433 .B64 2001
    635.9′647′0973—dc21

                    00-011727

A catalogue record for this book is available from the British Library.

The paper in this book meets the guidelines for permanence and durability of
the Committee on Production Guidelines for Book Longevity of the Council
on Library Resources.

10 9 8 7 6 5 4 3 2 1

PHOTOGRAPH BY SALLY ATKINS

This book is dedicated to the late

William A. Niering—

friend, scholar, inspiring teacher,

and passionate environmentalist—

who fought for the integrity of

the natural world

Contents

The Genesis of This Book

This book represents a collaboration between the faculty and students of the School of Forestry and Environmental Studies and the School of Art and Architecture at Yale University. "The American Lawn," a graduate seminar, was offered at Yale University during the spring semester of 1991. The following graduate student authors contributed to this book: Jon H. Connolly, Jennifer Greenfeld, Anne H. Harper, Lee Ann Jackson, William L. Kenny, Barbara Milton, John Petersen, Susan L. Pultz, Chris Rodstrom, Lisa Vernegaard, and Jennie M. Wood.

As environmentalists and educators, the authors of this volume have been concerned with environmental education for many years. Environmental problems are most often complex and difficult to understand. Teaching environmental studies in a coherent and integrative way is no easy task. In the fall of 1990, the first author proposed that the American lawn would be an excellent vehicle for teaching environmental principles. He was soon joined by his two colleagues, and the team designed a course for the spring of 1991, with the American lawn as its central theme. The goal of the course was truly unique—to compile a book-length manuscript on the topic "the American lawn: is it an environmental anachronism?" With the assistance of Dr. Joseph Miller, librarian and historian, the eleven graduate students examined and modified the chapter outlines proposed by the three author-instructors. Student teams were assigned to write each chapter. After reading the literature and interviewing lawn experts and knowledgeable faculty members, the teams wrote their first drafts. These were distributed to all participants and subjected to public and private criticism. The drafts were rewritten and the process was repeated. The present chapters are the work of the students, modified, substantially rewritten, and added to by the authors.

All proceeds from this book will be used to provide fellowships for students interested in ecology at the School of Forestry and Environmental Studies and at the School of Architecture at Yale University.

Acknowledgments

This book is a collaboration of many people and organizations. The authors would like to thank all who gave of their time, answered our many queries, and patiently assisted us.

This book could not have come into being without the persistence and assistance of the staff at Yale University Press. Jean E. Thomson Black, science editor, encouraged us from the beginning.

Anyone who has worked at Yale University knows of the invaluable resource provided by the library system and its staff. The late Joseph Miller of the Henry S. Graves Memorial Library at the School of Forestry and Environmental Studies provided expert guidance and readily responded to requests for assistance.

All the authors called upon the assistance of those who work with them. Audrey Anderson assisted with research. Carol Ziegler helped to produce countless drafts of the text. April Reese helped verify and update this edition's statistics.

Many individuals gave freely of their time. Our thanks go to Steven Beissinger, Christine Bormann, Dan Botkin, William Dest, Larry Forcier, Frank Golley, Morgan Grove, Jean-Marie Hartman, Gene Likens, Darrell Morrison, Carleton Ray, William Smith, Bill Welsh, and George Woodwell. Numerous individuals were interviewed and several in particular provided additional information and help: Mary and Bob Burks, Dave Harris, Steve Grout, Karen Williams, Kathy Garvin, Betsey Wright, Joseph Warner, Melody Hughes, Debbie Poor, Tara Pike, Ali Abbasi, and Larry Coffman. For helpful comments on the portions of the draft, we thank Dick Holmes, Pat Blum, Eric Stiles, and Pat Sutton.

The book's illustrations were produced by several talented artists and photographers. Those who worked closely with us include Lauren Brown, Karen Bussolini, Tony Casper, Alex Taylor, and Susan Hochgraf.

We contacted several organizations throughout the country while conducting our research. Special thanks go to Eliot C. and Beverly C. Roberts of the Lawn Institute (www.lawninstitute.com).

This book could not have been produced without the financial backing of the Mary Flagler Cary Charitable Trust and Edward A. Ames, trustee, whose quick and enthusiastic response encouraged us and confirmed the usefulness of the original book and this new edition.

Prologue to the Second Edition

Sustainability: What is it and how do we get there? We all know that with ever increasing human numbers and activity, we need to pay attention to the possibility that Earth's life support systems may be stressed. We know that we must search for and implement strategies that maintain life-support systems now and in the future.

This book is our primer toward sustainable lawn management. In the first edition, we focused on the lawn of our houses, its history, the pleasures and problems it gives us, its capture by the lawn-care industry, and the environmental and social costs of that capture. The lawn advocated by the lawn-care industry with its monoculture, chemicals, and waste of resources we called the Industrial Lawn. We proposed the Freedom Lawn, mowed when needed, free of pesticides and fertilizers, and often designed to reduce the proportion of the yard maintained as lawn. Our analysis of these lawn alternatives illustrates the possibility of a sounder relationship of our society to nature.

This new edition updates the original text and adds a new chapter tracing the diffusion throughout society of the concept of the Freedom Lawn since the publication of the first edition in 1993.

In our new material we examine how the concept of the Freedom Lawn has been accepted and adopted by many aspects of our society as we continue to realize the true costs of the Industrial Lawn in terms of aesthetics, environment, and economics.

We explore how concern about the Industrial Lawn has generated community organization and action leading to new landscape designs, fostered debate among educators about what is taught in the classroom and practiced in the management of the university campus, and influenced corporate thinking about management of their landscapes. We also probe the relationship between suburban developments with their Industrial Lawns and the negative effects they have on the hydrologic cycle and on regional landscapes. We urge a new approach to landscape design for the one million acres undergoing suburbanization each year. We show the potential connection between landscape development and negative effect on intra- and inter-continental bird migrations, which, in turn, can affect the ecology of huge regions. We report how citizens of a major bird stop-over point are working to diminish the negative effects of suburbanization on bird migrations. Finally we illustrate how some highway departments are introducing environmentally sound alternatives for the tens of millions of acres of rights-of-way under their management.

Americans, more aware of the pitfalls of the Industrial Lawn, are increasingly adopting environmentally sound landscape management designs and procedures. This new edition documents the progress made in the last decade. We hope that the Freedom Lawn will continue to spread across the United States, right up to the front door of the White House. We hope that the philosophy behind the Freedom Lawn will stimulate changes in lifestyles to achieve sustainability and to protect the health of the Earth's ecological systems upon which we all depend.

Prologue to the First Edition

Americans' attachment to the lawn is a long and fond one. A lawn is a gathering place for family, friends, and neighbors, a place where we engage in our favorite activities. In cities, it is a place of verdure, a refuge from crowds, traffic, and noise. The green blades feel good to the touch; the cut grass freshens the smell of the air. No other nation, except perhaps England, holds the lawn in such reverence. In passing through suburban neighborhoods where one landscaped lawn follows another, we can vividly see the pride Americans take in their lawns.

Our long love affair with the lawn has had its rough spots, but none more critical than its recent implication as another factor in the deterioration of the environment of the earth. What possible connection can there be between the lawn and the earth's biosphere? It is the purpose of this book to explore the numerous connections, to point out the many ways that we as lawn owners through our lawn management practices diminish in small but collectively significant ways local, regional, and global environments, and finally to suggest ways by which we can enjoy the many virtues of the lawn while reducing our impact on nature.

For many of us, the realization that our actions may be contributing to the deterioration of the planet is recent. About thirty years ago, signs of a steady decline in environmental quality became visible: polluted streams and rivers, smog-shrouded cities, and urban decay were everywhere. Scientists began to document more subtle effects: food chains contaminated with pesticides and air masses polluted with an extraordinary variety of wastes from the enormous engine that throbs at the heart of our society.

What seemed a patchwork of environmental problems, each specific to a particular region or country, began to coalesce into global phenomena: global warming caused by the accumulation of greenhouse gases in the atmosphere; large regions of the earth affected by acid rain with the potential for serious damage to streams, lakes, and forests; and chlorofluorocarbon pollutants that thin the stratospheric ozone layer with a resultant increase in biologically destructive ultraviolet light reaching the surface of the earth.

All this is beginning to reveal to us a global environmental deficit, the unanticipated consequence of humanity's alteration of the earth's atmosphere, water, soil, flora, fauna, and ecological systems. We are beginning to see that human activities may be disrupting the very life systems on which we all depend.

What place does the American lawn occupy in this scenario? Whether suburban backyard or city park, a gracious expanse or a tiny strip of green, the lawn is part of the earth's surface. It is the most commonplace landscape and the one most familiar to us. For the homeowner, the lawn is also our piece of the biosphere, and through it we communicate our concern about the environment of the earth, our greater yard.

To see the effects of ecological deterioration on our own lawns or in our own backyards is to have abstract ecological problems brought home. Issues of geochemical and hydrologic cycles, soil formation and species diversity seem abstractly important but beyond our reach. Ecological disasters that can contaminate the groundwater of whole regions of the Midwest or wipe out the tropical forests of a whole continent seem almost beyond our comprehension, let alone our ability to prevent. Yet, ecologically destructive practices that threaten the global environment are present in smaller landscapes, including our lawns. Understanding the dynamics of lawn ecology may bring to a human scale the meaning of ecological sustainability. By the same token, understanding this small ecological system may impel us to constructive action. We can begin to see that our weekend puttering on the lawn can mean caring for the planet. Fostering the development of an ethic of "environmental awareness" and exploring ways of implementing that ethic on the small piece of the environment entrusted to our care are the dual purposes of this volume.

To understand how our actions in managing the lawn might affect the biosphere, it is necessary to understand how the naturally occurring ecological systems, ecosystems that the lawn replaces, function. Forests, prairies, or other naturally occurring ecosystems are composed of large numbers of plants, animals, and microorganisms. Ecosystems function both above and below ground; they change with the seasons; but most of all they are systems powered by the sun (figure 1).

Using solar energy, ecosystems carry out an extraordinary array of processes. They store and recycle nutrients such as carbon and nitrogen, holding them within the system instead of releasing them to interconnected streams, lakes, and groundwater. They provide a home for the many species of organisms that live within their boundaries. Naturally occurring ecosystems make major contributions to the stability of the earth through their maintenance of air, water, and soil conditions favor-

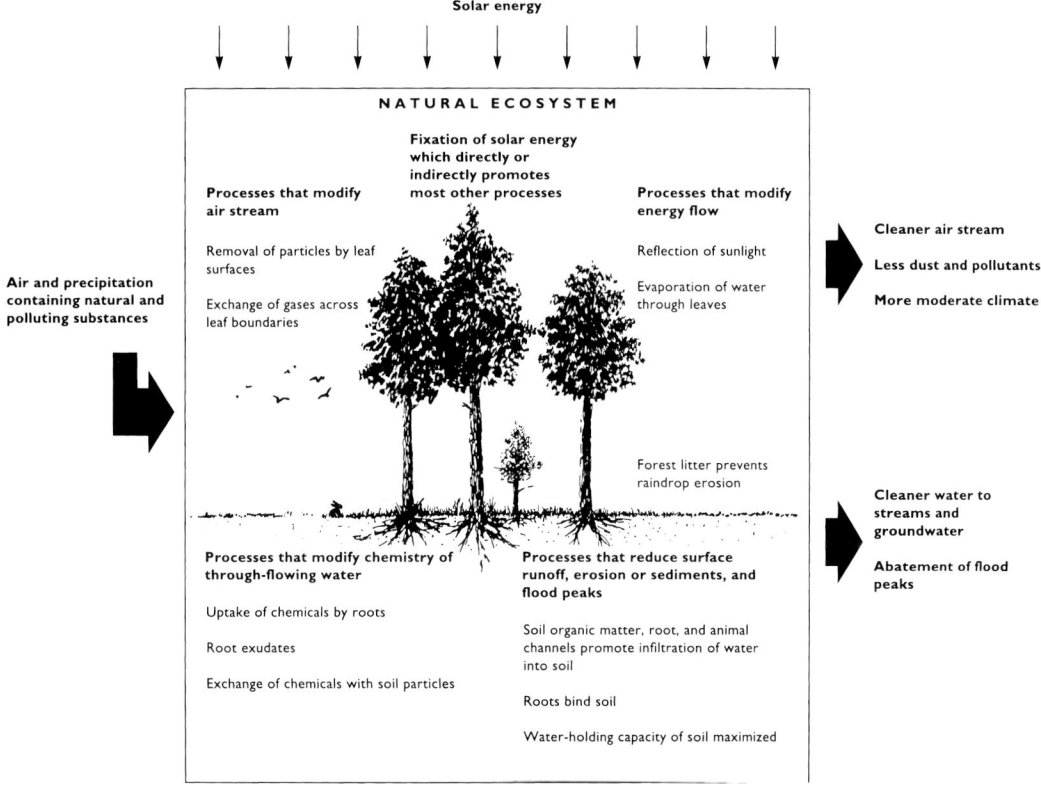

FIGURE 1. Using solar energy, naturally occurring ecosystems such as forests, prairies, and fields modify air and water in ways beneficial to human society. © 1985 American Institute of Biological Sciences.

able to human society everywhere on the planet. All of this is done using only the energy of the sun.

Human civilizations can affect the most remote ecosystems through air and water pollution, and many naturally occurring ecosystems have been drastically changed by urbanization, agriculture, and forestry. Many such changes are essential if our societies are to exist in their present form. Yet perhaps we are going too far in our manipulation of nature, so far as to be damaging the very life-support systems upon which we depend. Predictions of environmental catastrophes within our children's lifetimes are no longer considered absurd. The 1992 Earth Summit in Rio de Janeiro, Brazil, reflected the concerns of people throughout the world for the health of the planet. It is in this light that we wish to examine the American lawn.

The American lawn is a human-modified ecosystem, and many questions arise concerning its ecological function. Does it, like naturally occurring ecosystems, contribute to global ecological stability? Or is it part of the problem? Alternately, in designing and managing our lawns, have we departed too far from nature's plan? If so, can we modify our lawn care practices to mitigate damage to the environment?

We have all grown up with the American lawn, but we must reevaluate our attachment to the lawn in light of continuing evidence that human activity is disrupting the biosphere. To replace the eighteenth-century notions of the beautiful lawn landscape is to shape a new aesthetic to go with our new ecological ethic. These new visions of our landscapes, ecologically sound and aesthetically pleasing, might also guide the way we build our cities and communities, and, in fact, the way we conduct our lives.

1

Love of the Lawn

ANY STREET, USA

The lawn holds an important place in the American view of an ideal life. Over vast areas of small towns and suburbs a spreading green carpet forms the background for living. The well-kept lawn is not only beautiful in itself; it also provides the setting for the house. In these homes people feel in harmony with their well-tended plots of land. Coming home after work one might loosen a tie or kick off one's shoes and feel proud and relaxed without knowing why. In the spring new blades of grass emerge, creating a carpet of green that is a perfect backdrop for the beauty of crocus, daffodil, tulip, azalea, and rhododendron. In summer the world is lush with vegetation, the air perfumed by freshly cut grass. The homeowner is filled with a sense of well-being. Although the season, location, and resources and energy of the residents may vary, the ideal persists. The well-kept front lawns roll down the street, providing open space and beautiful vistas. In this ideal, the grass sward is as pure as possible, mowed two inches high, and free of dandelions and other insidious intruders (figures 2 and 3).

Behind these similar front lawns lie more varied backyards: some contain children's play equipment; others have patios, picnic tables, and barbecue grills; still others have gardens of vegetables or flowers. Here and there is an occasional pool. Evidence of unfinished tasks lies about: a pile of wood half stacked from last winter; a building project, perhaps lumber for a tree house, covered with plastic. Clothes hang on clotheslines. Under all of these activities the lawn rolls on, with bare spots marking the heavily used areas.

FIGURE 2. A suburban lawn. Photo: Lisa Vernegaard.

Whereas the front lawn is a bit like the parlor, the back lawn is more like the kitchen. The owner may be willing to share the front lawn for community functions, but privacy is what matters most in the backyard. You can step out on the back lawn, take a deep breath, and feel the sun on your face. It is just for you. In the front yard, a social obligation could arise at any moment, but in the backyard, you can set up a hammock, lie down, close your eyes, and relax. The backyard is yours alone.

WHY WE LOVE THE LAWN

Anyone who has ever played croquet or soccer on it or laid a blanket over it to listen to a summer concert or watch fireworks has no need to ask why Americans love the lawn. But if we look below the surface, our love of the lawn is more complicated. It involves aesthetics, economics, psychology, and especially history.

FIGURE 3. Connecticut Street, Litchfield, Connecticut. In suburbs and small towns throughout America, front lawns run together without interruption, giving a neighborhood a sense of unity and providing a source of community pride. Photo: Diana Balmori.

The lawn expresses a familiar aesthetic: its green expanse provides the framework for flower beds and shrubs, majestic trees, and our homes. Its predictable horizontality counters the verticality of trees, homes, and flowers while enhancing details of their surfaces. Shrubs and trees bordering the lawn's edge in turn complement the flat landscape they enclose.

The grass sward leads the eye of the viewer toward the horizon and into distant areas. Landscape features that begin in the foreground may be followed toward a distant and nebulous end point on the horizon. The lawn has been repeated over and over again, millions of times, yet it continues to fill us with delight and appreciation (figure 4).

Practical advantages should also be given their due. Some of the lawn's most beneficial functions are safety and health related. In some areas grass can serve as a firebreak, keeping wildfires at bay. Even before scientists discovered that turf can trap some pollutants and pollens that cause allergies, Walt Whitman called grass "the handkerchief of the Lord."[1] Lawns have great recreational value. They are home to baseball, croquet, touch football, badminton, and tag. Grass wears well and provides a cushioning effect that reduces injuries and makes walking, running, and jumping more comfortable (figure 5).

FIGURE 4. This streetscape illustrates the dominance of the lawn in suburbia. Our views of nature, the games we play, and our weekend chores are heavily influenced by this choice of landscape. Reproduced with permission from Yale School of Forestry and Environmental Studies, Center for Coastal and Watershed Systems.

The clean, cool, natural greenness of a beautiful lawn provides a pleasant environment. Its soft surface reduces glare and noise and conspires to dim, just a bit, the hustle and drive of our society. On any college campus, hundreds of students spread-eagled on every available patch of lawn bear witness to the comfort the lawn offers on the first warm days of spring.

Perhaps our love of the lawn contains elements that are both evolutionary and psychological. John Falk, who for years has been studying the human preference for grassed landscapes, sampled a diverse segment of the U.S. population to get a sense of their landscape preferences. Overwhelmingly, people favored short grass and scattered trees. Children especially preferred short grass.

Falk suggests that because humans evolved in the grassy, tree-sprinkled savannas of Africa, our modern preference for lawns and trees is an innate expression of our origins. The evidence for adapted predisposition is circumstantial at best, but Falk maintains that the connection is not unreasonable: "If we buy the notion of genetic vestige," he says, "then humans will find lawns a habitable, safe, and potentially supporting environment." Why does he think people love lawns? Falk laughs and replies, "Because we can't help ourselves."[2]

Some psychologists agree that humans have an innate preference for open spaces. Open spaces provide "legibility," an environment that is clear and easily understood, one where people are more likely to acquire information and less likely to get lost.[3]

FIGURE 5. Lawns are safe havens for kids of all ages. Ball games, tag, picnics, and naps are often our first experiences with nature. The songs of birds and the smell of newly mowed grass are two of the most relaxing sensations suburbanites enjoy. Photo: Karen Bussolini.

The lawn even has political overtones. Thomas Jefferson laid the groundwork for lawn ownership by advocating an order where "as few as possible shall be without a small portion of land."[4] Our second president, John Adams, proposed "the only possible way then of preserving the public virtue is to make the acquisition of land easy to every member of society: to make the dividing of land into small quantities that the multitudes may be possessed of landed estates."[5] Adams and Jefferson may not have been thinking of the lawn when they advocated universal ownership of small plots as the foundation for American democracy, but such ownership is indeed the basis for our millions of lawns.

Economics unquestionably plays a major role in our "love" of the lawn. A home is the cornerstone of many people's net worth—their primary asset. Great efforts are expended to maintain the home's value; because landscaping can add up to 15 percent of a home's worth, lawns contribute to resale value.[6]

FIGURE 6. The lawn of the medieval garden was full of small flowers and a mixture of grasses kept short by use. The medieval lawn, an early ancestor of the modern lawn, was known as the flowery meade. Upperrhenish painter about 1410, *Das Paradiesgärtlein,* Städelsches Kunstinstitut Frankfurt am Main. Photo: © Blauel Kunst-Dia.

HISTORY OF THE LAWN

For all of these reasons, it would seem—particularly if genetic preferences have any validity—that all cultures throughout history would have made lawns central to their public and domestic spaces. Yet this is not the case. Lawns arose primarily in Western civilization, and mainly in France and England, and it is the English part of its history that is particularly important to Americans' love of lawns.

In terms of human history, the lawn is not an old tradition. Its popularity began in eighteenth-century Europe, though its antecedents are deeply embedded in humankind's struggle to understand and control nature.

Medieval gardens were a form of environmental control: surrounded by walls, they provided a psychological sanctuary for human activity. Practical gardens provided a cultivated area where food and herbal medicines were grown, while in pleasure gardens, fruits and flowers invited lounging, dancing, and romance (figure 6).[7]

FIGURE 7. Productive and ornamental garden, Vaux le Vicomte, Seine-en-Marne, France. This seventeenth-century French garden has panels of grass as a background to the planted edges and to the perfect figures of fountains and parterres. The crushed stone paths, rather than the panels of grass, are for walking. Photo: DiaFrance.

Gardeners carefully used the limited space in these walled gardens, espaliering, dwarfing, and raising planters over bare rock at times. To save space, they often planted such crops as fruit trees in staggered rows, closely trimming the trees. The tight geometric composition of productive gardens behind walls may have been the forerunner of the geometric layout of agricultural fields and of the seventeenth-century French aristocratic gardens. As political power in France shifted from the individual walled fiefdoms to a centralized monarchy, the walls of these aristocratic gardens slowly came down (figure 7).

In the open French formal gardens of the seventeenth century, gardeners experimented with raised beds of grass shaped in ornate patterns. Behind these precise patterns lay a belief in a nature that worked with clocklike precision, and the image of a clock was a frequent metaphor for the hidden workings of the universe. Visitors strolled on walks of gravel woven through elevated grassy beds shaped in the precise yet graceful curves and arabesques of embroidery. These ornamental gardens were actually best viewed from the house, where guests could look down upon the complex patterns and be impressed by the presentation of order and human control over nature.[8]

Thus, French gardens did not reflect their surrounding environment but rather presented an abstract structuring of nature that became an art form. The garden was an area where human intelligence could reveal nature's hidden order.

Until the 1700s, English gardens closely imitated these formal gardens. In the eighteenth century, as views of nature began to shift, the English began to rival the French as landscapers. The English landscape, like its French counterpart, reflected political and cultural tenets that governed eighteenth-century English society. Gardens designed in England began to vary from the typical seventeenth-century French formal garden, developing a new aesthetic sprung from a major philosophical revision of the idea of nature.[9]

The invention of the sunken fence about 1690 allowed landscape gardeners to create the perception that the estate extended to the horizon unimpeded by fences. The French may have invented these structures, but the English used them to shape a new English landscape. These sunken fences ran like trenches, perpendicular to the line of view, and, when placed correctly, created an invisible barrier between the estate boundary and the cultivated or wild landscape on the other side. The exclamation of surprise people gave when they came upon these barriers unexpectedly is said to have earned them their name, *ha-ha* (figure 8). The deep, wide ha-ha kept grazing animals from wandering off the property and brought unimpeded distant views into the garden, giving the impression that the family estate extended to the horizon.[10]

FIGURE 8. A sunken fence, or ha-ha. Illustration by John James, 1712(?). Reproduced with permission from Anthony Huxley, *An Illustrated History of Gardening* (London: Paddington, 1978), 97.

English landscape gardeners carried this idea to an extreme that was limited only by money and site-specific constraints. The eighteenth-century English landscape historian Horace Walpole lauded William Kent (1685–1748), the preeminent English landscape gardener who made extensive use of the ha-ha fence.[11] According to Walpole, Kent "leaped the fence and saw that all nature was a garden."[12] Kent's designs created an image of space that exceeded the designs of his predecessors. He was inspired in part by the way eighteenth-century landscape painters were depicting nature. Painters, poets, and landscapers frequently made reference to a classical Arcadia, part of ancient Greece, invoked as a mythical reminder of a simple, pastoral life where humans lived in harmony with nature. In the late eighteenth century, an industrializing society again turned to the myth of Arcadia as an antidote to the urbanization and mechanization of life.[13]

This flourishing landscape art movement rode the coattails of a transformation in what was considered nature but what was in reality just another human-made landscape. Landscapers skillfully engineered compositions of nature, creating vistas out of rocks, trees, and water, erasing all marks of human activity to produce peaceful pictures of nature. As cities grew increasingly polluted and disease-ridden, landscapers transformed the countryside for the elite, often razing entire villages in order to create a view of nature—a nature presumably devoid of human intervention.[14] Taking center stage in this theater was the grass plant (see box). Using grass, landscapers blended estates into the property surrounding them, producing an illusion that their created landscapes flowed smoothly into nature.

Although William Kent and other landscape artists contributed to the establishment of the lawn as the essential component of the English landscape, it was Lancelot "Capability" Brown who brought the lawn to its full prominence.[15]

Capability Brown, a landscape artist who earned his nickname from his habit of discussing a particular site's "capabilities," would frequently obliterate the work of his

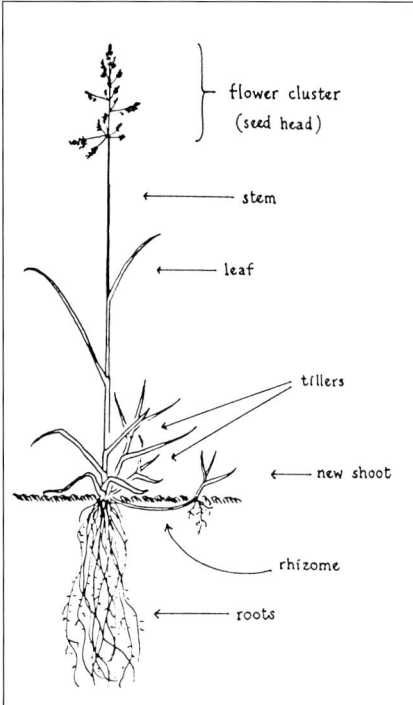

flower cluster (seed head)

stem

leaf

tillers

new shoot

rhizome

roots

Grass plants (figure 9) grow from a crown near the soil's surface and produce stems (composed of leaf bases) and blades (leaves). If mowing or grazing ceases, grass plants produce flowering stems that produce seeds. Grass invades surrounding areas by producing new plants from seeds or from the growing points, or by sending out rhizomes. Lawn grass is but one of the ways wild grasses have been domesticated; much of the world's food comes from domesticated grasses such as wheat, rice, barley, oats, rye, sugarcane, millet, bamboo, sorghum, corn, and pasture grasses. Although various grasses thrive in a range of environmental conditions, all share unique biological characteristics. Drawing © Lauren Brown.

predecessors, hiring gardeners to contour land into perfectly smooth surfaces on which sweeping lawns would be planted, thus giving landowners their own expansive vistas of "Nature." Brown's landscapes thrived as gardeners scythed, brushed, and swept their lush green surfaces, rolling the verdant carpets to smooth out irregularities.

Brown rarely created completely flat areas but, rather, shaped the ground into a concave or convex surface to focus an observer's view in a particular direction. Unlike Kent, Brown avoided creating painterly scenes that invoked images of Arcadia, preferring simpler open spaces and smooth lines.

Brown's penchant for planes of grass seems to have known no bounds. In the name of Nature, he destroyed yew hedges, displaced villages, dismantled houses, and chopped old avenues of mature trees to the ground. In Moor Park, Hertfordshire, he specified that an old garden be uprooted in order to plant numerous lawns that soon became famous all over England (figure 10). As Scottish landscaper Ian Hamilton Finlay has said of him: "Brown made water appear as Water, and lawn as Lawn."[16]

The success of Brown's landscapes became the keystone of a revolution in aesthetics that cemented the lawn as the great icon of late eighteenth-century British society. It cannot be overemphasized how closely Brown's success was tied to the

FIGURE 10. Moor Park, one of the most cited of Capability Brown's landscape designs, is described by Thomas Jefferson as "thirty acres of lawn." Richard Wilson, *Moor Park*. © Marquess of Zetland. Reproduced in Roger Turner, *Capability Brown and the Eighteenth-Century English Landscape* (New York: Rizzoli, 1985), 85.

English climate. England is a country of mild winters, moderate temperatures, and high humidity—all conditions favorable to the growth of the grasses he planted so abundantly. The lawn soon became a symbol whose roots reached across the Atlantic to different, less hospitable environments.

THE HISTORY OF THE LAWN IN THE NEW WORLD

When the British colonized the New World, they brought with them their ideas about nature. Cultivated, grazed, and developed over many centuries, the island of Great Britain had no unaltered landscapes left. British landscape design that "improved" a natural landscape in reality only added new artificial dimensions to an existing landscape already highly modified by humans. In the New World, colonists were ill prepared to cope with seemingly infinite expanses of wild land inhabited by unfamiliar plants, animals, and people.

The British also brought a strong preference for an English landscape to the New World. Although new climates and different vegetation and soils challenged English aesthetics and ways of life, the immigrants clung to practices and ideals of their cultural past. Vast expanses of wild land ready for farming only fueled this image.

Grass played a central role in English agriculture because it sustained sheep and cattle. William Wood, an English traveler, warned settlers coming to New England in the 1630s to seek places where there would be enough grass to feed cattle. In general, grazing land was scarce; domesticated animals quickly ran out of grass, and forestland then had to be cleared to create new pastures. Seed brought from Europe was used, and soon such European plants as bluegrass and white clover, which were adapted to the harsh requirements of pastoralism, began to take over wherever cattle grazed. By 1640 a regular market in European seed existed in Rhode Island, and within one or two generations these plants had become so common that settlers regarded them as native. During the eighteenth century, European pasture plants including timothy and fowl-meadow grass and the legumes red clover and alfalfa were common on pastures throughout the colonies.[17] Thus the pasture, a mixture of grasses, legumes, and other plants, became a common landscape feature in much of colonial America. Pasture was not only a rural feature but was also seen in many villages and towns, where it clothed the village common.

One of the many travelers who brought the English landscape to the United States was Thomas Jefferson. Jefferson visited and admired some of the most important English gardens, and he was clearly struck by the pastoral image of a classical building set in a field of green. "Moor Park . . . the lawn about thirty acres" is recorded in his notebook.[18] He later incorporated this vision in the design of his home Monticello, where he combined European classical precedents of villa design, English landscape ideas centered around the lawn, and American ideals of independent citizen-farmers. The view from his estate began at the mansion, flowed through a lawned garden and woods to the view of productive farm landscape, and ended with the wild, unsubdued scenery on the fringes of the estate.

About this time he created a place for the education of his citizen-farmers: the University of Virginia. Here the buildings were organized in the shape of a U surrounding a tiered lawn. In Jefferson's original design, the viewer was to look out on nature in the form of a botanical garden, which was never built, and beyond to the distant hills. The whole Jeffersonian complex of the University of Virginia

FIGURE 11. The Jeffersonian complex at the University of Virginia centers around a tiered lawn and is known to thousands of graduates as "The Lawn." Photo: Diana Balmori.

is called "The Lawn," its name based on the beautiful lawned space it contains (figure 11).

Lawns were not uncommon in America at the turn of the eighteenth century, but with few exceptions they were kept to a minimum, the grass cut by hand scythes or kept short by grazing animals. At Mount Vernon, George Washington created a great tree-lined swath of grass sweeping down to the Potomac beyond the ha-ha that separated it from the smaller lawn around the house. Cutting this expanse of grass was a cause of concern until he hit on the happy idea of using browsing deer to keep it short.[19] Much more common than lawns were treeless, shrubless, rather unkempt weedy properties whose front yards, especially in the South, were tidy patches of swept bare ground with occasional planting beds or shrubs.

THE SUBURBAN LAWN

When did the mowed lawn take such firm root among the American conscious-ness? According to Kenneth B. Jackson's classic work *The Crabgrass Frontier: The Sub-urbanization of the United States,* the mowed lawn appeared in the mid-nineteenth cen-

FIGURE 12. A New England green. Photo: Diana Balmori.

tury along with the American suburb.[20] The Industrial Revolution had transformed American cities, and a new middle class began to seek homes outside urban areas. Parks, cemeteries, and suburban cottages were recommended for their aesthetic, moral, and recreational benefits as well as their contribution to physical health. Builders landscaped green areas around houses to satisfy clients who believed that green plants created reservoirs of clean air and healthy home environments. By 1870, detached housing had emerged as the suburban style of choice, with drawings typically depicting an isolated structure surrounded by a yard.

The lawn, carrying the English connotations of nature with it, became a symbol of prestige in the nineteenth-century suburbs. Similarly, centers of towns in New England, the old "commons," which had been the setting for such useful activities as rope making, hay growing, military drills, and town fairs in the eighteenth century, were transformed in the nineteenth century from bare stamped earth, cultivated fields, or cemetery grounds into lawned and treed parks, now called "greens." With this grassy transformation, most of the economic functions of the commons were shifted to other places. At least one American writer, J. B. Jackson, has condemned this transformation of true "commons" into "greens" (figure 12).[21]

The upkeep of green spaces required intensive labor, and early suburbanites depended on hired labor or on sheep or deer to keep things tidy. The advent of the lawn

FIGURE 13. Budding's invention of the lawn mower quickly replaced the scythes and brooms once used to maintain the lawn. William Wollett, *West Wycombe,* 1770 (detail). Copyright © The British Museum. Early advertisement for the lawn mower reproduced with permission from University of Reading, Institute of Agricultural History and Museum of English Rural Life.

as the common people's art form would be made possible by technology: in 1830, the Englishman Edwin Budding invented the lawn mower (figure 13). As these machines became available over the following decades, modest householders could own these mowers and keep their lawns tidily cut without the help of gardeners or flocks of sheep. When applying to the British Patent Office, Budding claimed: "Country gentlemen will find in using my machine an amusing, useful and healthful exercise."[22] Jane Loudon heralded the invention of the lawn mower as "a substitute for mowing with the scythe . . . particularly adapted for amateurs . . . but it is proper to observe that many gardeners are prejudiced against it."[23] Forty years after the lawn mower's introduction, Samuel Orchart Beeton's *Dictionary of Everyday Gardening* described the use of mowers as "too well known to need description. . . . These useful machines are fast supplanting the scythe both on large and small lawns."[24]

Finally the well-manicured lawn was within easy reach of average citizens. Whether they had an acre or a tiny patch of land, they could tend it themselves and create whatever image they chose for their surroundings.

In the 1840s, two Americans, Alexander Jackson Davis and Andrew Jackson Downing, vastly extended the popularity of the small detached houses of England in the United States, publishing highly successful books that contained numerous illustrations of bungalows surrounded by lawns and gardens. Downing was convinced that the environment directly affected behavior. His *Treatise on the Theory and Practice of Landscape Gardening, Adapted to North America* (1841) discussed the princi-

FIGURE 14. Andrew Jackson Downing's landscapes, as shown here, incorporated trees, lawns, and gravel walks. From Andrew Jackson Downing, *A Treatise on the Theory and Practice of Landscape Gardening* (1841), 1. Reprinted 1977 (9th ed., 1875) by Theoprastus Publishers, Sakonnet, R.I.

ples of landscape design founded in natural scenery.[25] This book, widely read, encouraged Downing to write several more, most with the theme of encouraging Americans to improve their surroundings, though his ideas were all based on English landscaping principles. A contemporary commented that "the value of Downing's books has been great, not because of their technical excellence, for they are very poor in quality, but because they are full of life and interest. It is the man not the architect that wins the popular ear, and compels his readers to allow that the subject is entertaining and enjoyable."[26] From its publication in 1841 until the end of the century, Downing's *Treatise* remained the average homeowner's standard reference. In it, the lawn is the unifying theme (figure 14).

Separateness, Downing pointed out, had become essential to the character of the suburb. As Kenneth Jackson writes, "Although visually open to the street, the lawn was a barrier, a kind of verdant moat separating the household from the threats and temptations of the city. It served as a means of transition from the public street to the very private house. . . . The sweeping lawn helped civilize the wild vista beyond and provided a carpet for new outdoor activities such as croquet, a lawn game imported from England in the 1860s, tennis and social gatherings" (figure 15).[27]

FIGURE 15. The tennis party. By the second half of the nineteenth century, lawn games became a feature of middle-class suburban homes. Sir John Lavery, *The Tennis Party,* 1885 (detail). City of Aberdeen Art Gallery and Museums Collections. By courtesy of Felix Rosenstiel's Widow & Son, Ltd., London, on behalf of the Estate of Sir John Lavery.

In New York and other large cities, zoning laws made their appearance and twenty-five-foot setbacks from the street became standard; these distances seemed monumental compared to the old row houses that abutted directly on the street.[28] On the city outskirts, legal covenants requiring structures to be set back from the street by a minimum number of feet were written into many property deeds from the 1880s on. "The idealization of the home as a kind of Edenic retreat . . . where the family could focus inward . . . led to an emphasis on the garden and lawn," writes Jackson.[29]

The lawn became the symbol of suburbia, championed by Frederick Law Olmsted, the famous landscaper of New York City's Central Park and many suburbs, who viewed the lawn as a sort of community parkland. For Olmsted, the front lawn of a house in a suburb unified the residential composition as one neighborhood, giving a sense of ampleness, greenness, and community.[30]

The curvilinear layout of American residential streets, with houses set well back

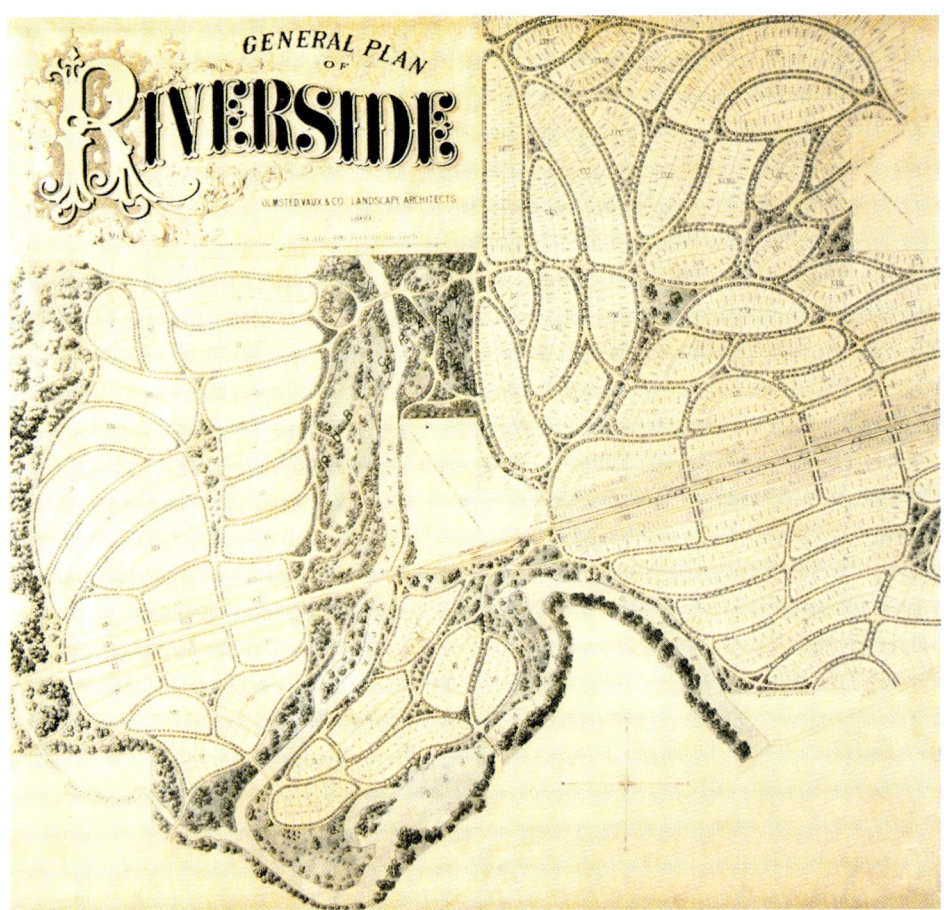

FIGURE 16. Plan for Riverside, Illinois, by Olmsted and Vaux, 1869. Winding streets with continuous front yards of lawns and trees defined the character of the new American suburbs. Curving streets and green lawns declared that the homes were "in the country." Olmsted, Vaux and Co., 1869 Oak Forest/Riverside Plan, reproduced in *Journal of the Society of Architectural Historians* 42, no. 2 (May 1983): 156.

from the road behind front lawns with informal plantings of trees and shrubs, a uniquely American residential form, was first proposed by and built for an industrialist by Andrew Jackson Davis in his suburb Llewellyn, twelve miles west of Manhattan.[31] Downing, Olmsted, and others popularized this layout. Olmsted and his partner Calvert Vaux's plan for Riverside near Chicago has become an archetype of this American suburb (figure 16).[32]

The lawn continues to have strong philosophical advocates today, although there is a shift in locale. William Whyte, sociologist and urban advocate, sees the lawn as important to modern American cities. In his book *City: Rediscovering the Center,* he writes: "A salute to grass is in order. It is a wonderfully adaptable substance, and while it is not the most comfortable seating, it is fine for napping, sunbathing, picnicking, and Frisbee throwing. Like movable chairs, it also has the great advantage of offering people the widest possible choice of seating arrangements. . . . Grass offers a psychological benefit as well. A patch of green gives us a refreshing counter to granite and concrete, and when people are asked what they would like to see in a park, trees and grass are usually at the top of the list."[33]

THE LAWN'S CONQUEST OF NORTH AMERICA

The lawn, since its invasion of the eastern seaboard nearly three centuries ago, has spread to every corner of North America. This is a bit hard to comprehend, because the lawn arose in the mild, moist climate of England, whereas North America is a continent with many harsh climates and an enormous diversity in vegetation and soils. Climates range from the extreme winter cold of Canada and the northern United States to the intensely hot and drought-prone summers of the South, from the extremely wet Northwest Coast, to the wet East Coast, to the extremely dry Southwest. These climatic patterns were reflected in the great range of natural vegetation the first explorers encountered when they painstakingly boated and walked over the surface of this vast continent. Evergreen forests crossed the continent in the north; deciduous forests covered the great stretch of land from the Mississippi eastward; native grasslands, not English pasture grasses, swept northward from the Gulf of Mexico, west of the Mississippi, and east of the Rockies into Canada; and in the Southwest large areas of desert vegetation occupied lowlands that ran well into Mexico (figure 17). Most of these vegetation types and climates were not in the least hospitable to the English lawn. Nevertheless, the same continuous sward of green front lawn joins block after block, whether on Pennsylvania Avenue in Urbana, Illinois, or Ocean Front Boulevard in Santa Monica, California. Just how could the mild-mannered lawn conquer these often hostile environments?

The adaptation of the lawn as the green mantle that unified the suburbs was initially made possible by the lawn mower. Budding's invention was followed by many others and paralleled by an explosion in agricultural knowledge and technology. Farmers and agronomists learned about plant nutrition, about fertilizers and how to produce them, about plant disease and insect depredations of plants and how

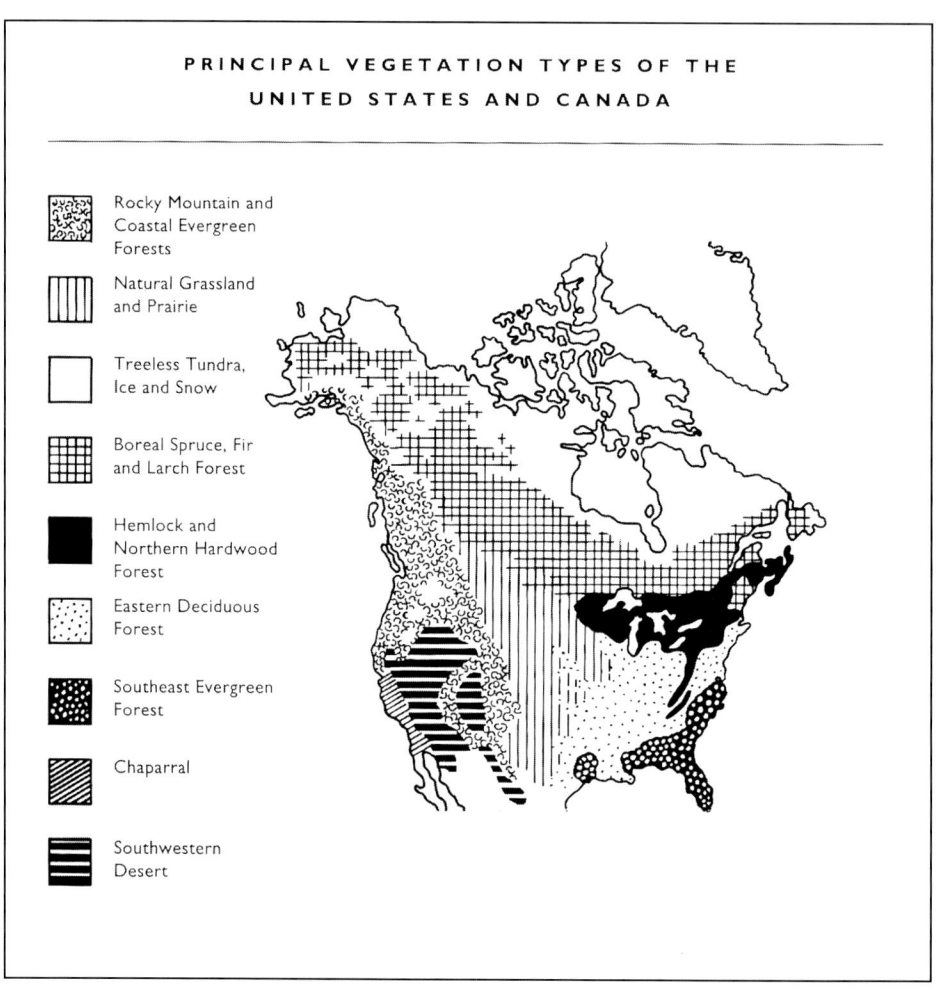

FIGURE 17. Principal types of vegetation in the United States and Canada. Adapted from Henry J. Oosting, *A Study of Plant Communities: An Introduction to Plant Ecology,* 2d ed. (San Francisco: W. H. Freeman, 1956), 271.

to combat them, about water requirements of plants and how to meet them with irrigation, and, perhaps most important, how to breed plants better adapted to particular environments.

Technology soon found its way into the development of the lawn care industry and through entrepreneurialism became relatively inexpensive and available to homeowners through retail outlets and professional lawn care companies. Grass seed companies developed varieties especially for the noncommercial market. Chemical companies responded to the needs of small-scale purchasers. Sod companies made the lawn shippable and made possible instant lawns. And thus the lawn was inexorably commercialized and through entrepreneurialism was able to overcome naturally occurring environmental barriers. Not only because it was possible but also because it was preferred and liked by a mass audience conditioned by a long cultural history, the lawn became the landscape of choice from Kennebunkport, Maine, to Las Vegas, Nevada.

But there was a price to pay.

2
Questioning the Lawn

Millions of Americans love their lawns and are satisfied with a rich sward of pure grass; but others have lawns that depart from this standard. These Americans have begun to question the purpose of a lawn and its social, environmental, and economic costs. Many are breaking with suburban orthodoxy and reconsidering the management and design of the space around their homes. In essence, they challenge the monolithic view of the American lawn.

We begin by describing the nonconventional lawns of homeowners in Maryland, Connecticut, and Georgia. Examining their dissent helps us understand what the American lawn is and what it may become. In the remainder of this chapter, we explore the rise in environmental awareness that has led to growing opposition to the traditional lawn.

WALTER AND NANCY STEWART OF POTOMAC, MARYLAND

Walter Stewart did not look forward to cutting the lawn, but when the weekend came around, he knew that it had to be done.[1] One day everything changed: the lawn mower failed to start. What better excuse than a broken mower to placate one's conscience! As time went by, the mower remained broken and the grass grew longer and longer; gradually it developed into a meadow. After a time, the Stewarts concluded that nothing except convention dictated that they must have a front yard of mowed grass. "People may think we are lazy, but who cares." This is precisely what their neighbors first thought, but soon their tolerance for this eccentric behavior turned to impatience. The Stewarts were informed that the neighborhood was not pleased with their landscape design; the neighbors insisted that a conventional lawn was more appropriate. The Stewarts responded with a lengthy letter explaining why they chose to let their lawn grow into a meadow. They proposed that a meadow provided more than a lawn could ever hope to provide: a haven for animals, food for countless organisms, privacy for the homeowner, and simply a more "natural" environment. They also noted that maintaining a conventional lawn was a waste of their time and energy, because it takes considerable energy to keep a lawn from turning into a meadow in Maryland.

Walter and Nancy Stewart received a citation from Montgomery County stating that their lawn was in violation of the twelve-inch rule and mandating that the lawn be cut within ten days. Any lawn that exceeded twelve inches in height was considered a municipal health risk. The Stewarts defended their choice of a meadow on ecological grounds and challenged the validity of the county's health risk ordinance. In the end, the county changed its regulations, and the Stewarts were allowed to keep their meadow.

MICHAEL POLLAN OF CORNWALL, CONNECTICUT

Michael Pollan is a homeowner who espouses a new sense of nature. He has chosen to deviate further and further from the green carpets of uniform height that conventionally connect homes in the suburbs. When Pollan first bought his home, he faithfully trimmed his lawn. Although this was not his favorite pastime, he did enjoy being outdoors and caring for his grass. From endless weeks of mowing, he came to know his yard intimately. He knew when bumps were coming, where the mower would feel the strain of an especially thick patch of grass, and when he was only five minutes from completion. He recognized that crabgrass grew in harsh areas and clover preferred depressions where water tended to collect.

With time the regimen of mowing lost its appeal and other aspects of his yard became far more fascinating. Pollan developed a particular liking for gardening, for growing vegetables and fruits. With time, fruits and vegetables became a significant part of his landscape design. Fruit trees were planted in his front yard, and by trial and error he learned the ecological requirements of various kinds of vegetables. There was a subtle pleasure in finding the right balance of water, nutrients, and light that would enhance the plants' growth. All of these pursuits he greatly preferred to mowing. One does not learn much from the grass that one mows. Even the word *mow* conjures up images of conquest and submission to indiscriminate assault. In *Second Nature,* an account of his experiences in the garden, Pollan remarks: "Gardening, as compared to lawn care, tutors us in nature's ways, fostering an ethic of give-and-take with respect to the land. Gardens instruct us in the particularities of place. They lessen our dependence on distant sources of energy, technology, [and] food."[2]

And so Michael Pollan continues to extend his gardens. His turf grows smaller and smaller; he has bordered his yard with a hedge, breaking the continuous carpet from home to home. And he has a half-acre meadow of black-eyed Susans and oxeye daisies that yield beautiful flowers from early summer to the time of killing frost.

MURRAY AND ANN BLUM OF ATHENS, GEORGIA

When Murray and Ann Blum built their house in Athens, Georgia, they insisted that the developer leave all of the trees already growing on the property.[3] Between these relatively mature trees, the Blums added sycamores, a Chinese walnut, and French locusts to provide beauty throughout the year. Murray Blum's lifelong interest in insects led them also to add flowering shrubs to provide dependable daily resources for insects for much of the growing season.

When the woody plants matured, the yard looked more and more "natural." Plants and animals found their way into the mix that the Blums provided. As the interactions between fauna and flora multiplied, the Blums realized that they were "giving something back to nature." Their front yard had become a part of nature again (figure 18).

"So we let things grow. We've let the ground covers work out their territories. The recent invasion of *Vinca* has provided the elements of a major battle as it approaches the English ivy." Nature has not been left totally to its own devices, however. Murray has great fun finding plants, especially wildflowers, to add.

The neighbors have not been happy with all of these activities. The Blums' yard does not fit in; it interrupts the green continuity of the street's front lawns. To appease the neighbors and to legitimize this one-acre irregularity, Athens posted a sign on a large tree declaring the Blums' property a one-acre bird sanctuary.

"Grass cultivation is energy-intensive, consuming vast amounts of resources, e.g., fertilizer, biocides, and water," says Murray Blum. "Grass seems to us an expensive anachronism in a world where people are starving. For us, the beauty of wildflowers and their pollinators far surpass the monotony of lawns. Our yard is no place to play baseball, but that is a small price to pay for the medley of trees,

FIGURE 18. Murray Blum in his front yard in Athens, Georgia. On May 6, 1992, the *Atlanta Journal* featured a story about the Blums and their yard. Photo: © William Berry, 1992. Reprinted with permission from the *Atlanta Journal-Constitution*.

shrubs, thickets, leaves, and flowers and for the omnipresent birds, bees, and bugs."

Starting with a desire to protect the trees growing on their newly purchased lot, the Blums have given their yard back to nature and at the same time have reduced their dependence on scarce resources. In their own front yard they are able to express their aesthetic preferences and their concern to promote harmony between humans and nature.

JOEL MEISEL OF PROSPECT, CONNECTICUT

Joel Meisel's dissent developed gradually, much as Michael Pollan's did.[4] Meisel noticed that wildflowers were continually invading his lawn. These were not simply dandelions or plantains but meadow flowers. As a scientist concerned with plants, Meisel's appreciation went beyond simple delight in beauty. *Krigia virginica,* a small, yellow-flowered plant, had its own natural history, one which he understood. He enjoyed the fact that these plants had found their way to his front yard and found himself spending idle moments contemplating the marvel of their growth, survival, and spread. He continued to mow his lawn, but he avoided the patches of wildflowers; soon his lawn had become a patchwork of lawn and flowers.

His neighbors were concerned about his increasing deviation from the perfect lawn. Inquiries were made to determine if something was wrong. To the more curious questioners, Meisel explained his interest in the little reproductive miracles that were taking place in his lawn. He found that some neighbors were impressed with his knowledge of what was going on in his lawn and thus respectful of his dissent. Soon other people in the neighborhood were growing wildflowers, although not in the "lawn." Meisel continues to mow around hawkweed, mullein, and oxeye daisies, and he and a next-door neighbor share a wildflower garden between their homes.

WHY DO SOME PEOPLE WISH TO CHANGE THE LAWN?

The examples cited above could be multiplied manyfold, from Arizonans who prefer a patch of desert to a carpet of green to North Carolinians who value a long-leaf pine forest more than a lawn. What message is there in this dissent? To answer this question we need to explore four elements of varying importance: time, changing aesthetics, people's desire to experiment, and environmental consciousness.

Many people clearly dislike the burden that constant lawn care places upon them. Time is often in short supply today. In the minds of many, lawn care can be a demanding taskmaster. The inexorable routine of weekly mowing, especially under the impetus of added fertilizers and irrigation, can become a boring and time-consuming routine. Some lawn owners find comfort in developing lawn management strategies that demand less time and energy. Others may hire a lawn care company that, for a price, relieves the owner of most responsibility.

In a very direct way, the experiences of the Stewarts, Michael Pollan, the Blums, and Joel Meisel reflect the beginnings of a change in landscape aesthetics. The eighteenth-century vision of a distant nature has lost its strength. In its stead, a sense of kinship and curiosity about living things is growing, and there are attempts to represent this more inclusive vision in the lawn and garden, for this is still where most daily contact with other living things takes place. In time, these efforts taken in conjunction with artistic visions will shape this new sensitivity into a different aesthetic image of the landscape.

As a people, Americans are explorative! Our history is filled with tinkering, innovation, and invention. The lawn offers an arena for experimentation. One of the most alluring aspects of "lawn tinkering" involves tinkering not with technology but with nature. The Stewarts became fascinated with a natural process called plant succession by which their lawn became a meadow through naturally occurring processes. Michael Pollan became interested in the potential of his land to produce things other than lawn grass, and Joel Meisel decided to use part of his lawn to study the behavior of wildflowers. The Blums saw in their yard an opportunity to give something back to nature. It seems clear that the interplay between imagination and exploration plays a large part in the dissent from the green carpet lawn. This impetus for innovation is aided by the examples we see around us, so the more people who innovate, the more examples we will all have.

THE RISE OF ENVIRONMENTAL CONSCIOUSNESS

A final reason for dissent from conventional concepts of the lawn is the rise of environmental consciousness. The following overview touches only the surface of a deeply philosophical and complex subject.

For much of human history, there was great fear of the unknown. The unknown was everything "out there." Indeed, nature had much that could thwart and humble humans: wild beasts, floods, earthquakes, plagues, wildfires, crop disasters, sickness, unexpected frosts, droughts, locust swarms, erosion, hurricanes, and many other natural events. Succor was centered on relationships between people, because these formed a bulwark against nature's unpredictability. Not unexpectedly, this perception of the relationship between people and their animate and inanimate surroundings led to a philosophy of "man apart from nature." Under this paradigm, humans viewed nature not only with a degree of fear but also as something to be manipulated in what were perceived to be the best interests of humanity. The philosophy of the eighteenth-century landscape designers reflected this view. For the upper classes, the pleasure garden and the lawn were considered human buffers against wild nature.

At about the time the lawn came into fashion, Europeans had gone far toward subduing nature. The demands of an increased population had changed Europe into a largely human-controlled landscape. Wilderness as we think of it was essentially gone. European culture was also beginning to feel the effects of the Industrial Revolution, however. In this world of machines, of crowded and disease-ridden industrial cities full of chimneys belching black smoke, the beautiful image of serene, grass-carpeted landscapes in rural areas unmarred by urbanization became a social ideal. The rural landscape, even though it was itself a product of human manipulation, was seen as nature when juxtaposed with the grim view of the human-created urban environment. The natural world of trees, mountains, pastures, grazing cattle, streams and rivers, and rocks was contrasted to the urbanized and industrialized works of humankind.

As northern Europeans continued to emigrate to North America, they carried with them this eighteenth-century philosophy. These new emigrants viewed the great expanses of wilderness through the eyes of Europeans whose landscapes had long been under human control; they looked through the prism of "man apart from nature." Wilderness was considered an impediment, a source of danger from Indians, wild animals, and the unknown. Wilderness needed to be cleared as quickly as possible to minimize these inherent dangers and to convert it to purposes useful to people. Nature was to be subdued and made to serve humanity.

Although the idea of "man apart from nature" has a history that stretches through the millennia, it reached its zenith in the twentieth century. The extraordinary accomplishments of the industrial age, with its fantastic advances in science, medicine, space exploration, engineering, computing, agricultural and industrial production, transportation, and communication, have led many to believe that the work of men and women can solve any problem. As the old navy saying goes, "The improbable we do immediately, the impossible takes a little longer." Powerful individuals in the fields of science, technology, finance, economics, industry, and politics firmly believe in the "technological fix": that human ingenuity, technology, and organization can do anything, that humankind knows no bounds. This kind of thinking, so common today, represents the ultimate in "man apart from nature."

DOUBTS ABOUT "PEOPLE APART FROM NATURE"

The idea that people are above the laws of nature has long been questioned. More than four hundred years ago, Sir Francis Bacon observed that "nature, to be commanded, must be obeyed." This aphorism has direct bearing on how we design and manage our lawns. For if we are *not* above the laws of nature, some of what we do to the lawn may change naturally occurring processes in such a way that nature's response will be inimicable to humankind's long-term best interests. In other words, we cannot indiscriminately manipulate our lawns without in some measure diminishing our local, regional, and global environment. This is perhaps the major reason why many people are troubled by the conventional view of the lawn as a lush green carpet of Kentucky bluegrass.

A great body of scientific knowledge has developed over the centuries that lends strong support to the idea that humans are not unique, at least in terms of the natural processes that govern the earth and the stars. Today science accepts the existence of a chemical basis for life. This chemistry underlies how we become ill, how we get better, how medicines work, how plants make food, and how organisms decompose. It is believed that every process in the biological world can be described in terms of chemical reactions. Although there is still a vast amount to learn, biochemists working with a huge range of organisms have demonstrated that many chemical processes are common to all organisms, including humans.

Inviolable forces are at work in the atoms that make up our bodies, and forces unimaginable to most of us are at work in the universe. The physics and chemistry of humans are not unique; our bodies are composed of matter that obeys describable laws in a universe so vast that humanity is nothing but a speck.

Other notions of human uniqueness were shaken with the publication of Charles Darwin's *Origin of Species* in 1859.[5] Darwin's ideas dramatically changed the way people saw themselves in the natural world. Today many people have come to accept that humanity shares a common lineage with other primates and with all other forms of life. We are not separate and distinct. Whales, elephants, mice, and humans are just different products of time and evolutionary processes.

Ecologists and other scientists who study whole forests, lakes, oceans, and the atmosphere have demonstrated a wide array of processes that are essential to maintaining life on the earth. These essential processes maintain the composition of the atmosphere, underlie the constant cycling of nutrients between living and dead matter, maintain the soil, and govern the cycling of water. All life depends on them.

The scientific evidence is overwhelming that humans are part of the fabric of nature and cannot be separated from it. We are living, breathing organisms subject to gravity, chemical bonds, energy flows, and nutrient cycles. We have a common lineage with other life-forms with whom we share common biological processes, and like other organisms we depend on naturally occurring processes to maintain conditions favorable to life. The new motto should be that the only things in this world you have to do are to get born, die, pay taxes, and *abide by nature.*

Environmental thinking challenges the idea that humans are above nature and that nature can be thought of simply as a commodity to be bought, sold, altered, or destroyed (figure 19). Environmentalists see all life as part of nature and believe that our manipulations of the earth's surface should be carried out with the deepest respect for nature lest nature respond with an irrevocable altering of the very earthly conditions upon which humankind is dependent for survival.

The roots of environmental thought are deep, and books about it fill many shelves. To catch the historical flavor of its evolution, it is useful to touch on the contributions of a few of its key thinkers. In 1851, Henry David Thoreau wrote:

> At present, in this vicinity, the best part of the land is not private property; the landscape is not owned, and the walker enjoys comparative freedom. But possibly the day will come when it will be partitioned off into so-called pleasure-grounds, in which a few will take a narrow and exclusive pleasure only,—when fences shall be multiplied, and man-traps and other engines invented to confine men to the *public* road, and walking over the surface of God's earth shall be construed to mean trespassing on some gentleman's grounds. . . . Hope and the future for me are not in lawns and cultivated fields, not in towns and cities, but in the impervious and quaking swamps. . . . Give me the ocean, the desert or the wilderness![6]

Thoreau's remarks were as spiritually oriented as they were foresighted. Although he wrote these words a century and a half ago, Thoreau expresses respect for nature and captures America's twenty-first-century concern to protect nature and pristine environments.

In his timeless book *Man and Nature; or Physical Geography as Modified by Human Action,* George Perkins Marsh in 1864 described the danger humanity faced if people continued to raze the forests of the world. The issues he set forth are still pertinent today and will remain so into the foreseeable future. Marsh chronicled the strong correlation between the destruction of woodland resources and the collapse of human empires and wrote scathingly about the land management practices of his day:

> Man is everywhere a disturbing agent. Wherever he plants his foot, the harmonies of nature are turned to discords. The proportions and accommodations which insured the stability of existing arrangements are overthrown. Indigenous vegetable and animal species are extirpated, and supplanted by others of foreign origin, spontaneous production is forbidden or restricted, and the face of the earth is either laid bare or covered with a new and reluctant growth of vegetable forms,

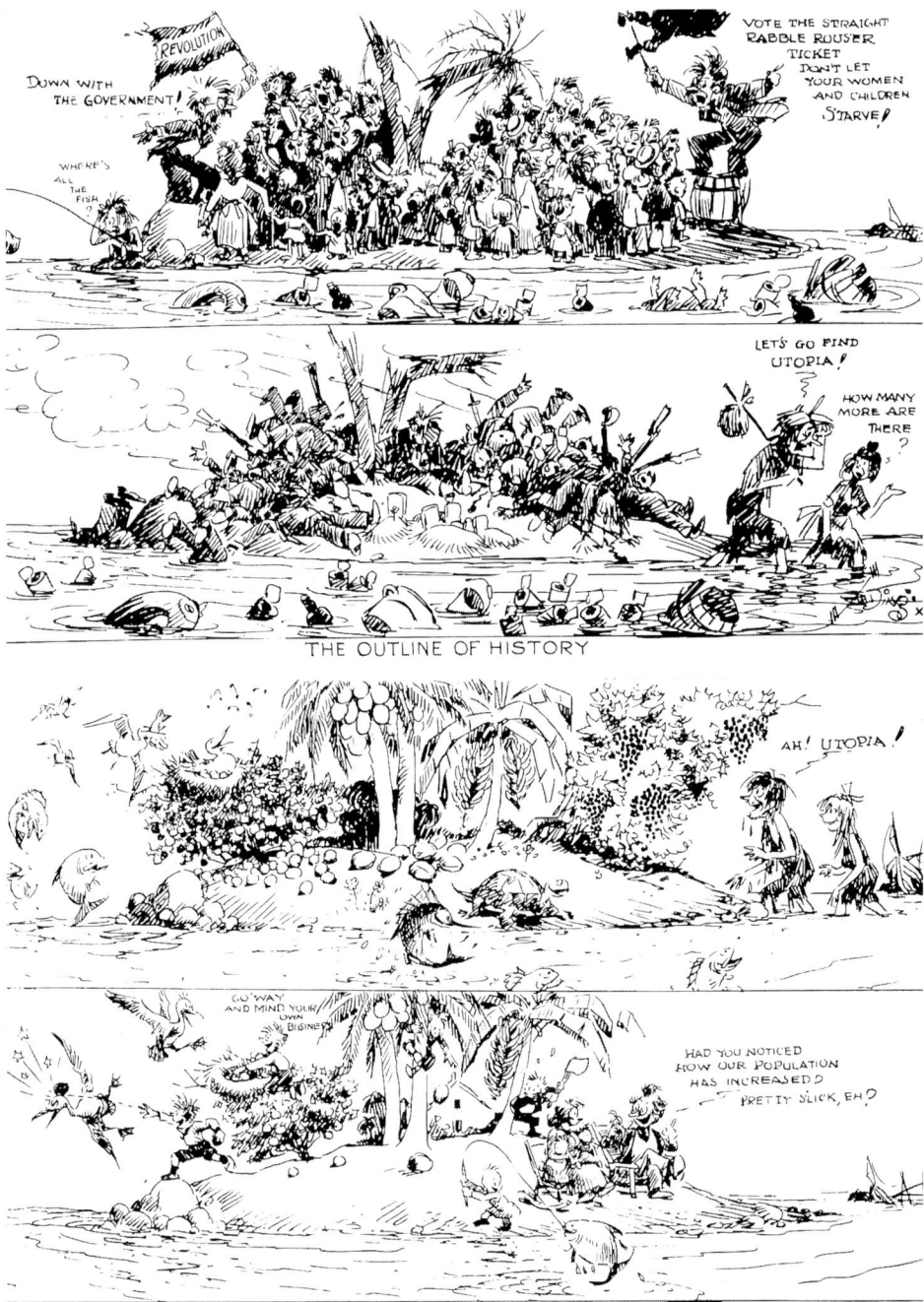

FIGURE 19. Ding Darling, famous political cartoonist and early environmentalist, captures the widely held belief that humans could destroy some part of the earth and simply move on to another in his 1936 cartoon of utopia. The cartoon is dedicated to Paul Sears, eminent ecologist of the mid-twentieth century and author of the renowned book on the great North American drought of the 1930s, *Deserts on the March*. Cartoon property of F. Herbert Bormann. © J. N. "Ding" Darling Foundation.

and with alien tribes of animal life. These intentional changes and substitutions constitute, indeed, great revolutions; but vast as is their magnitude and importance, they are . . . insignificant in comparison with the contingent and unsought results which have flowed from them.[7]

Marsh clearly and forcefully saw the unanticipated side effects of nineteenth-century technology, a problem that has grown to enormous proportions in today's world.

In 1948, William Vogt wrote *Road to Survival*.[8] Among his provocative ideas were the notions that technology makes the world effectively smaller and that localized acts can affect the entire planet. Vogt challenged the concept of limitless space, which held that humans could destroy one area and move to another, or, in a modern context, that we could discharge our hazardous wastes knowing that almost infinite dilution would render them harmless, or that we could store them in out-of-the-way places where they were set off from human society. Vogt challenged the widely accepted notion that more growth was always better and that technological advancement was an inherent and unequivocal good.

One year later, in 1949, Aldo Leopold's book *A Sand County Almanac* described humanity's role in light of all life around us.[9] The book articulated the connection between preserving wilderness and preserving all nature, including ourselves. Arguing from a strong foundation in both science and ethics, Leopold contended that we were dooming ourselves by disrupting nature. Most humans imagine that they are sustained by economy and industry. Leopold did not deny this fact, but he pointed out that what sustains economy and industry is all living things.[10] In essence, he told us that too narrow a focus in our economic pursuits can result in unanticipated responses by nature that could prove dangerous to human society.

In the early 1960s, in her stunning book *Silent Spring*, Rachel Carson (figure 20) made the American public keenly aware that our personal health was being compromised by our general neglect of ecological connections.[11] In her indictment of the use and overuse of chemical pesticides in America, she argued that our culture had headed willy-nilly into the age of chemicals without considering the totality of nature; processes within nature were working to undo the very things we had hoped to achieve using pesticides, thereby causing large-scale imbalances. In our manipulations of nature, she urged us to seek to use the processes and inviolable realities of nature to our advantage and to avoid the use of pesticides.

The public's initial response to *Silent Spring* was an instinctively corporeal one, a concern for human health, rather than an understanding of the broader implications of her work: the need to know our place in ecology. Yet this response seems natural when one reads some of the passages from *Silent Spring*:

FIGURE 20. Rachel Carson was among the first scientists to alert the American public about the unappreciated and dangerous side effects of the chemical age. Photo: © Erich Hartmann/Magnum Photos.

In one of the most tragic cases of endrin poisoning there was no apparent carelessness; efforts had been made to take precautions apparently considered adequate. A year-old child had been taken by his American parents to live in Venezuela. There were cockroaches in the house to which they moved, and after a few days a spray containing endrin was used. The baby and the small family dog were taken out of the house before the spraying was done about nine o'clock one morning. After the spraying the floors were washed. The baby and dog were returned to the house in midafternoon. An hour or so later the dog vomited, went into convulsions, and died. At 10 P.M. on the evening of the same day the baby also vomited, went into convulsions, . . . and [ultimately] became little more than a vegetable.[12]

Here is a story all American parents could relate to. This is a story not of occupational exposure but of the high price paid for innocent ignorance of the effects of these chemicals. The message to the public was clear: to ignore the potential dangers of these chemicals was to place one's own children and pets in peril in one's own home.

Rachel Carson also demonstrated that it was not just acute incidents that we had to fear. The accumulation of small doses of chemicals over long periods of time was another source of danger. After World War II, pesticides were being sprayed everywhere for every conceivable purpose. Carson made Americans question the ultimate fate of these chemicals. When the smell was gone, were the chemicals themselves gone? Carson demonstrated that these chemical poisons were often incredibly effective because they could persist in the environment for a long time. These chemicals were designed for use in a natural setting that had a limited ability to decompose them. If these chemicals persisted in the environment, it could mean that there was abundant opportunity for human exposure at unlikely times in unsuspected places. The focus came back to human health in a very personal way. As Carson wrote:

> The contamination of our world is not alone a matter of mass spraying. Indeed, for most of us this is of less importance than the innumerable small-scale exposures to which we are subjected day by day, year after year. Like the constant dripping of water that in turn wears away the hardest stone, this birth-to-death contact with dangerous chemicals may in the end prove disastrous. Each of these recurrent exposures, no matter how slight, contributes to the progressive buildup of chemicals in our bodies and so to cumulative poisoning. Probably no person is immune to contact with this spreading contamination unless he lives in the most isolated situation imaginable.[13]

Indeed, these dangerous chemicals can be found as close to home as one's own lawn. Carson observed that "those [lawn-keepers] who fail to make wide use of this array of lethal sprays and dusts are by implication remiss, for almost every newspaper's gardening page . . . take[s] their use for granted."[14]

These prophets touched on many points. George Perkins Marsh emphasized that unanticipated side effects from forest destruction acting through ecological pathways would have strong negative effects on human societies. William Vogt stressed how technology can shrink the globe, making space less of a solution to our environmental problems. Although we are sustained by our economy and industry, Aldo Leopold told us that these things were, in turn, sustained by nature. To injure nature was to injure ourselves. Finally Rachel Carson focused on the biogeochemical pathways of nature and how chemical contamination of these pathways can injure humans as well as other organisms and damage the ecological systems that provide the basis for society's health and well-being. Henry David Thoreau pointed out that nature has spiritual and moral values in addition to practical values. Although we have not explored this track to any degree, surely these values have a central place in any debate about nature and our use of it.

The gateway opened by *Silent Spring* was followed by a flood of environmental literature in the late sixties and early seventies, among them such books as Barry Commoner's *The Closing Circle—Nature, Man, and Technology* (1971) and Paul Ehrlich's *The Population Bomb: Population Control or Race to Oblivion* (1968).[15] That flood has continued to this day.

One of the most important groups of environmental writers today can be found at the Worldwatch Institute under the leadership of Lester R. Brown. Many consider the institute's annual publication, *State of the World,* as important and influential as *Silent Spring* and *Sand County Almanac.*[16] Published in thirty-two languages, *State of the World* documents the evidence, which is accumulating at an alarming rate, that unanticipated side effects of humankind's actions are causing the severe deterioration of natural systems. To quote Lester Brown, "Anyone who regularly reads scientific journals has to be concerned with the earth's changing physical condition. Each major indicator shows a deterioration in natural systems: forests are shrink-

What happens to fallout and pesticides in our environment?
Some of it decays - the rest ACCUMULATES.

FIGURE 21. In 1966, George Woodwell, W. M. Malcolm, and R. H. Whittaker published this series of cartoons in a small booklet entitled "A-Bombs, Bugbombs, and Us." The booklet, built on the theme raised by Rachel Carson, emphasized that chemicals and radioactive materials moving through the food chain affected not only plants and animals but humans at the end of the chain. Woodwell played an important role in raising environmental consciousness and was one of the founders of the important environmental organization the Natural Resources Defense Council. From Brookhaven National Laboratories and the U.S. Department of Energy. (Figure continues on pages 42–44.)

It gets into plants, then plant-eaters, then meat-eaters.

All along this food chain it's CONCENTRATED because animals eat many times their own weight in food.

FIGURE 21 *(continued)*

So the last guy in the chain gets quite a load.

WE'RE LAST GUYS.

Poisons kill other last guys, too,
making nature simpler and less stable.

FIGURE 21 *(continued)*

Imagine a simple food chain: *posies → sheep → people*

In nature: *SIMPLICITY = INSTABILITY*

What's more, pests get immune to poisons, and we must forever beef up the poisons or find new ones.

FIGURE 21 *(continued)*

ing, deserts are expanding, croplands are losing topsoil, the stratospheric ozone layer continues to thin, greenhouse gases are accumulating, the number of plant and animal species is diminishing, air pollution has reached health-threatening levels in hundreds of cities, and damage from acid rain can be seen on every continent."[17] Collectively the evidence strongly indicates that humans are very much a part of nature and not in some way exempt from nature's rules.

That the concerns expressed by the environmental thinkers are shared by many people today can be seen in the large growth of such environmental groups as the Environmental Defense Fund, the Natural Resources Defense Council, the Sierra Club, the National Audubon Society, the National Wildlife Federation, the Wilderness Society, Friends of the Earth, and Greenpeace. These organizations have millions of members, whose awareness of threats to the environment is supported by ever more sophisticated scientific research and motivated by a concern for their children and future generations. A new public view of nature is emerging that recognizes the complex integrated relationships that bind all living things. We are becoming increasingly aware of the cumulative and potentially disastrous impact of imprudent use of resources on ecological systems and on all forms of life, including humans.

Walter and Nancy Stewart endured some ridicule when they allowed their lawn to become a meadow. They made this change for reasons that involve a sense of beauty, but they were also influenced by the *new* sense of nature handed to us by science, writings, cultural events, and the evolution of that phenomenon we call environmentalism: the sense that we are not independent of nature but reliant upon it for our continued existence and that humanity can have a profound influence on the course that nature takes on this planet. The Blums are consciously being "environmentalists" when they encourage native plants and animals to grow and live in their yard. It seems likely that Michael Pollan and Joel Meisel were also influenced by a new sense of nature when they decided to dissent from the conventional lush green lawn.

This new understanding of nature, our role in it, and our ability to influence its course has Americans thinking in new ways. Many of us are desperately concerned about the global environmental problems that face us. We wonder, just as desperately, what we can do to alleviate these problems that often seem so vast they leave us feeling helpless and powerless. Yet there are many things we can do to tackle serious global environmental problems. Some opportunities for change lie just beyond our front and back doors: the lawn. By careful thought we can express our concern for nature by adopting an ecological approach to designing and caring for our lawn. We can strive to preserve some of the lawn's cherished values while diminishing some of its contributions to local, regional, and global environmental crises. For many people, the lawn is a clear opportunity to "think globally, but act locally."[18]

3

The Economic Juggernaut

Making the decision to have something other than a pure grass, continuously green, manicured lawn involves bucking a universally accepted view of what a yard should contain. In sports parlance, it is Westwood High School against the Chicago Bears. Here we explore the idea that this unequal match is a result of the powerful economic forces in a market economy. Let's begin by asking, What exactly is a lawn?

THE FREEDOM LAWN

Lawn, n., a stretch of grass-covered land, esp. one closely mowed, as near a house or in a park.

Broadly interpreted, this definition does not prohibit the presence of other plants. In practice, it is easy to find mowed lawns that include many kinds of plants other than grass plants. A few common to northeastern lawns are: dandelion, violet, bluet, spurrey, chickweed, chrysanthemum, brown-eyed Susan, partridgeberry, Canada mayflower, various clovers, plantain, evening primrose, various rushes, and wood rush, as well as grasses not usually associated with the well-manicured lawn, such as broomsedge, sweet vernal grass, timothy, quack grass, oat grass, crabgrass, and foxtail grass. All of these plants can coexist quite nicely with grasses considered to be *the* lawn grasses, such as bluegrass, ryegrass, and fescue. All of these potential inhabitants can tolerate mowing because they can keep sufficient energy-fixing apparatus below the level of the mowing blade. Some will remain green and healthy but never produce flowers because their erect flowering stalks are cut off by the mower.

We call this the "Freedom Lawn," for it permits all kinds of plants to exist in the only way they know how—by growing (figure 22). The Freedom Lawn results from an interaction of naturally occurring processes and the selective effects of lawn mowing.

The Freedom Lawn is continually bombarded by seeds from nearby herbs, shrubs, and trees. Some of these may find an open space and germinate, producing a new plant that, if it can tolerate the whirring blade, will become part of the lawn. One of the most interesting things about the process of plant establishment is that it does not occur equally everywhere. In most lawns there are subtle variations: moist depressions, droughty tops of mounds, somewhat cooler north-facing slopes or warmer and drier south-facing slopes, some areas always exposed to full sun and some partially shaded by nearby trees. These little variations in location alter the conditions of growth and survival of plants; some plants are better adapted to particular conditions than others. Thus the flora is far from uniform; the plants that compose the lawn tend to vary in response to the microenvironment and competition from other plants, that is, moist spots will have an abundance of species that thrive in moist environments, whereas species that best compete for scarce water will be found in droughty spots. Thus, the plant arrangement in the Freedom Lawn is well designed by the interaction between mowing and local ecology. The plants that succeed here do so without artificial intervention and collectively produce a green cover that is adapted to the peculiarities of place. In other words, with relatively little effort it is possible to have a green lawn adapted to the site.

An important part of the adaptability of the Freedom Lawn is its built-in capacity to tolerate various kinds of stress. Because of the diversity of plants, for example, a particular disease or insect is unlikely to wipe out the lawn. Some plants might suffer, but others will prosper and expand into areas vacated by the affected plants. The same is true during prolonged droughts; the lawn might turn brown, but it is unlikely that it will die. Many plants have special mechanisms for surviving drought and are able to resume active growth and return greenness to the lawn when the moisture returns.

THE INDUSTRIAL LAWN

In the sweeping parklike arcades of suburbia, where continuous lawns follow the graceful curves of residential roads, the Freedom Lawn is seldom found. "Freedom" in the lawn violates possibly all of the cardinal principles underlying the ac-

FIGURE 22. The Industrial Lawn (above) and the Freedom Lawn (right). Photos: Lisa Vernegaard.

ceptable suburban lawn—principles that any hardware store or lawn care center shouts at you by means of their lawn care products. These products include grass seed specially designed for sun, shade, or in between; boxes and sacks of fertilizer of different formulations; squares of grass sod for instant lawns; bottles, cans, and sacks of insecticides and herbicides; hoses, sprinklers, and irrigation-control devices; edgers and spreaders; mulching and sweeping machines; aerators and rollers; how-to books of lawn care; and mowers ranging from the inexpensive push mower hidden in a corner to the cadillac four-wheel-drive rider mower featured on a pedestal. All of these support a different definition of the lawn.

Lawn, n., a stretch of grass-covered land, especially near a house or in a park, that is regularly and closely mowed, continuously green, and, to the greatest possible degree, free of weeds and pests.

This is the lawn your neighbors expect to see on their trip home after battling the evening rush hour; the lawn created by what has come to be known as the lawn care industry. In fact, we might think of it as the "Industrial Lawn" (figure 22; see also figure 23).

The Industrial Lawn rests on four basic principles of design and management: (1) it is composed of grass species only; (2) it is free of weeds and other pests; (3) insofar as possible it is continuously green; and (4) it is regularly mowed to a low,

even height. In contrast to the Freedom Lawn, which in the hands of a relaxed lawn owner requires little other than an occasional mowing, the Industrial Lawn depends utterly on the expenditure of considerable money, time, and energy. The Industrial Lawn is not attuned to the peculiarities of place. Like energy-intensive agriculture it ignores microclimates and species diversity and substitutes technology for natural processes. It has the added virtue, at least in terms of the lawn care industry, of never being completely attainable. There is always some new and necessary bit of technology, some new finding on fertilizers, some modification of pesticides or new variety of grass required to move toward the ideal or, in more competitive terms, to keep up with the neighbors.

How did such a powerful paradigm arise? The existence of an estimated $30 billion a year turfgrass industry might raise suspicions of a smoke-filled room with CEOs plotting the manipulation of American homeowners and rubbing their hands together in anticipation of the increasing flood of money as the message of technological dependence is spread by wily lawn care promoters.[1] Of course this is not the case. Instead the Industrial Lawn seems to have developed from a marriage of many forces: the long history of our love of the lawn coupled with technological advancements that made modern agriculture possible, agricultural corporations searching for new outlets for their products, and skilled marketers fighting for market shares.

FIGURE 23. An all-grass, pest-free, continuously green, and constantly mowed Industrial Lawn.
Photo: Peter Miller.

Decades ago entrepreneurs identified the lawn as a potential market. They adapted existing agricultural technology to lawn management and design and through research developed new technology. They participated in the development of private and publicly supported organizations that promoted the care and culture of turf. Using marketing skills and advertising they were able to sell their products and gradually shape the concept of the lawn to meet their desire for increased profit. Thus arose the Industrial Lawn, by some standards a good thing, for it underlay a diverse industry that created jobs and income and contributed to the gross national product.

The American lawn in its industrial form presents a powerful symbol that explains, in part, the resistance that Meisel, the Stewarts, and the Blums met from their neighbors when they decided to change their lawns. Manicured green lawns unite the front yards of millions of suburban houses on both sides of the street to create an expansive and unifying parklike setting. Michael Pollan captures the primitive force of this deeply ingrained view of the lawn: "To stand in the way of such a powerful current is not easily done. Since we have traditionally eschewed fences and hedges in America, the suburban vista can be marred by the negligence—or dissent—of a single property owner. This is why lawn care is regarded as such an important civic responsibility in the suburbs, and why, as I learned as a child, the majority will not tolerate the laggard. . . . That subtle yet unmistakable frontier, where the crew-cut lawn rubs up against a shaggy one, is a scar on the face of suburbia, . . . an intolerable hint of trouble in paradise."[2]

THE GROWTH OF THE LAWN CARE MARKET

Like a king's crown, any patch of lawn represents riches imported from the far corners of the kingdom. Fertilizers, pesticides, irrigation systems, water, tools, gas and oil, labor, and the very seed itself are all convened for the purpose of growing greener grass for a longer part of the year. Lawn and lawn care demands have created a large and steady market within each of these related industries. It is the scale and arrangement of these different components that describe the industry's regional characteristics, the variety of distinct lawn markets, and overall changes in the direction of the industry.

One hundred years ago most Americans would have laughed at the idea of a lawn industry. Pastures were common, and special measures to grow grass domestically would have seemed unusual. Today the total area in turfgrass is about 27.6 million acres, an area about the size of Pennsylvania.[3] Grass is the selected ground cover for athletic fields, parks, playgrounds, highway verges, cemeteries, golf courses, schools, and homes. Seventy-six percent of the total, or 21 million acres, is in home lawns.

How did a large and complex industry evolve out of an individual's pursuit of a greener lawn? The lawn industry got its start in 1901 when Congress allotted seventeen thousand dollars to study the "best native and foreign grass species . . . for turfing lawns and pleasure grounds."[4] In 1920, the industry gained real momentum when the Greens Section of the United States Golf Association lobbied and won the support of the Department of Agriculture for a program researching grass species suitable for greens and fairways. Today grass research centers are found in most states, and courses on turf and its maintenance are offered by agricultural universities. The private and public infrastructure promoting the lawn and its maintenance is huge. Homeowners are caught in a Sisyphean cycle of seeking a greener and greener lawn, a cycle curbed only by their willingness to pay.

In marketing, perception is everything. In a 1981 study of eight hundred lawn owners in the Piedmont and Coastal Plain regions of Virginia, 80 percent thought that their lawn was average or below and were not satisfied with its present condition.[5] There will always be a lawn that is greener than yours, be it on a golf course, in a photograph, or covering a neighbor's front yard. This sets the stage for an infinitely expanding array of lawn care products and accessories.

The lawn care industry has become a growth industry constantly searching for new rationales for the expansion of lawns or the expansion of sales of lawn care products. The Professional Lawn Care Association of America (PLCAA), which promotes lawn and turf, now advertises not just improved property values resulting from greener lawns but increased health benefits as well: "Healthy turf means healthy lives." The PLCAA brochure "ABC's of Lawn and Turf Benefits" includes as advantages groundwater enhancement, buffering sports injuries, discouraging littering, providing a link with nature, and being generally therapeutic for humans.[6]

The Lawn Institute states that a fifty-by-fifty-foot lawn produces enough oxygen to sustain a family of four, an assertion that we show to be inaccurate in Chapter 4. The industry lobbied heavily in support of the 1990 congressional farm bill that advocated tree and lawn planting in urban areas to combat noise, dust, and global warming. Clearly these ecologically and socially sound ideas will also produce economic benefits for the lawn care industry: the more grass is planted, the more lawn care products are sold. Segments of the industry have advocated changes favorable to a sounder ecology: in 1990 the industry launched a national campaign to promote grasscycling: "today's turf tomorrow's earth." Several months of press releases and other publicity advocated leaving the cut grass on the lawn after mowing to increase the return of nutrients to the soil and to reduce yard waste entering landfills. Since

1994, more than a hundred companies and organizations have signed on to the Environmental Protection Agency's Pesticide Environmental Stewardship Program, designed to reduce commercial use of pesticides.

Industry organizations such as the Turf Resource Center and the Outdoor Power Equipment Institute are making some legitimate attempts to promote environmentally sound lawn care, but the lawn industry as a whole is also capitalizing on marketing strategies that appeal to Americans' new environmental consciousness. One industry publication, *Lawn News,* titled an article "Sell the Environment to Increase Your Profits."[7]

THE DIMENSIONS OF THE INDUSTRY

The word *grass* is believed to derive from the old Aryan word *ghra,* also the root of *grain, green,* and *grow.* Grass was elemental, the "general herbage" that sustained life. "All flesh is as grass. And all the glory of man is as the flower of the grass" (Isa. 40.6). Today healthy, green growing grass has become more of a symbol of success and plenty rather than the annual crop on which life depends. Grass grown for lawn is not a food, not a medicine. It has become for many a luxury item basic to being a satisfied and responsible homeowner.

Because it is a nonessential product, most of the documentation concerning the size and complexity of the lawn industry comes from organizations with a vested interest in the business. The *Annual State of the Industry Report,* prepared by the editors of *Landscape Management,* for example, is based solely on the experience of the industry's membership, a self-selected group of corporations. Professional magazines generally publish information on topics of interest to their readers, such as pest management, golf course design, sports turf, and the latest grass seed hybrids. The magazine titles call out to their readership: *American Lawn Applicator, The Greenmaster, Golf Course Management, Outdoor Power Equipment, Grounds Maintenance,* and *Turf News.* These magazines publish few concrete statistics. Data detailing national trends and regional markets are high-priced trade statistics and the very stuff of competition. The profits of these corporations and professional lawn care services are based on knowing the product and the market better than the competition.

Much of the available documentation pertaining to lawns derives from the National Gardening Association's annual survey conducted by the Gallup Organization. This unaligned poll is the most objective description of the industry available. The 1998–99 survey indicates that 49 million American households (47 percent of all U.S. households) are engaged in do-it-yourself lawn care. Far more Americans participate in lawn care (mowing, seeding, weeding, and so forth) than in other yard activities, including flower gardening (40 million), shrub care (26 million), insect control (23 million), and landscaping (23 million). American lawns cover an estimated 21 million acres with an average size of about one-third of an acre.[8]

These home lawns, green witnesses to countless American Saturdays, collectively occupy more acreage than any agricultural crop and account for the lion's share of the lawn industry's profits. Lawns range from highly manicured "estate" lawns to relatively low-maintenance lawns; in essence, from the Industrial Lawn to the Freedom Lawn. Lawns at the industrial end require considerable annual investment, whereas those at the freedom end require only a minimum investment. Millions of lawn owners are in the middle range, each managing their average one-third acre by fighting sporadic battles against poor soil fertility, drought, insects, and incursions of children and wildlife. Although low-maintenance lawns constitute a considerable acreage, major profits for the lawn care industry come from the upper and middle maintenance levels.

In 1998, the National Gardening Association estimated annual retail sales of residential lawn care products and equipment at $8.5 billion, up 34 percent from 1997. Annual turf and lawn maintenance is altogether a $30 billion industry. One might wonder how lawn care expenditures compare to expenditures made to raise agricultural crops. Although the comparison is a complex one, Augusta Goldin has reported that, per acre, it costs more to maintain the Industrial Lawn than it does to raise corn, rice, or sugarcane.[9] In the context of one's own backyard, $30 billion is a difficult number to fathom. The lawn industry represents less than 1 percent of the U.S. gross domestic product, but this percentage in no way adequately reflects the industry's power to determine the structure of our landscapes and to influence our psyches and attitudes toward our environment.

THE COMPONENTS OF THE INDUSTRY

The four principles underlying the Industrial Lawn—all grass, weed and pest free, continuously green, and closely mowed—determine the components of the lawn industry. To establish a lawn, one requires a source of grass seed. After germination, a grass plant uses the sun's energy to produce plant biomass: leaves, roots, stem, flowers, and fruits. The amount of biomass is determined in large measure by the amount of water and such nutrients as nitrogen, phosphorus, and potassium present and available in the soil. The grass plant's habit of growing up from below, probably an evolutionary adaptation to grazing animals, makes it ideal for mowing but also creates a recurrent need for mowing, raking, and other maintenance tasks. To encourage a lawn to be continuously green, fertilizers (nutrients) and water may be added. Some insects and fungi view a lush green lawn as the centerpiece of a huge bacchanalian feast. Thwarting the desires of these pests requires the application of insecticides and fungicides. To inhibit the growth of weeds, herbicides can be applied. The endless nature of this cycle is clearly evident: water and nutrients promote growth; herbicides lessen competition from nongrass weeds; rapid growth requires frequent mowing; pesticides inhibit not only grass predators but also organisms that decompose clippings; to achieve the desired appearance, clippings require removal; the removal of clippings requires fertilizers to replace lost nutrients. The cycle is music to the ears of the lawn care industry.

Seed

Although seed accounts for a small percentage of the annual costs of maintaining a home lawn, that small investment for each lawn adds up to an important segment of the billion-dollar seed industry. Until World War II, grass seed, like most other agricultural seeds, was traded as a commodity. Different varieties of grass supplied by many individual growers and sellers filled the market. As the seed market has become increasingly industrialized, producers have had to look for new ways to hold on to their market share. In recent years, producers have transformed grass seed into a specialty product in an effort to appear different and better to the perplexed consumer. A look at the packages of grass seed at the local hardware store gives an idea of current approaches to marketing problems. Producers now advertise seed mixtures for heavy traffic, low maintenance, deep shade, and full sun. In July 2000, the *New York Times* reported on research by the Scotts Company, the world's largest maker of lawn and turf products, to develop new grass strains genetically altered to withstand application of potent weed killers and remain green and apparently healthy.[10] According to the report, one new strain called "low mow" has been designed to grow at a slower pace, while other strains could be drought resistant or able to flourish in winter. To survive, however, all strains will need the application of potent weed killers also produced by Scotts. The user, once committed to this path, will thus need to apply weed killers continuously to maintain these genetically altered grasses. The article reports that serious questions have been raised by critics about the effect of the potential movement of genes from the genetically altered plants into native species, but no mention is made of potential ecosystem effects on soil biota, nutrient cycling, the hydrologic cycle, and the like, nor are overall energy and social costs described. The report stresses that company executives expect to reap enormous profits.

These genetically altered grass-weed killer systems would seem to be the ultimate expression of the industrial lawn. An alternative that seed companies might explore is the development and promotion of a more species-diverse lawn—what we are calling the Freedom Lawn. The value of such an alternative is discussed in Chapter 6.

Fertilizer

Like other components of lawn care, fertilizer is a product originally developed for agriculture. As agriculture markets stabilize or stagnate, the fertilizer industry has sought growth through expansion into the lawn care market. By weight only 5 to 10 percent of the fertilizer sold in the United States is purchased to fertilize lawns, but this market accounts for 25 percent of the industry's profits. In fact there is little difference between lawn fertilizers and agricultural fertilizers other than the size of the package and the type of marketing.[11]

The fertilizer industry's marketing strategy involves convincing the homeowner of the need for repeated applications of fertilizer to the lawn. The concept of the continuously green lawn and the removal of grass clippings from the lawn contribute to this need. Removing grass clippings from the lawn is analogous to removing the ears of corn or silage in corn culture; fertilizers become necessary to make up for nutrients lost through removal.

The $17 billion fertilizer industry (connected to both agriculture and lawn care) is highly dependent on mineral resources and abundant energy supplies.[12] Fertilizers for lawns are a mix of nitrogen, phosphorus, and potassium, the familiar N-P-K, just as agricultural fertilizers are. The production process is entirely dependent on fossil fuels, particularly oil and natural gas. Natural gas is the main ingredient in the production of nitrogen fertilizers and the main reason why oil companies with natural gas holdings own fertilizer companies. Although fossil fuels are not a direct component of the production of phosphorus and potassium, they are heavily used in mining, processing, and transporting these nutrients to the market.

Growing public concern over the environmental impact of heavy fertilization is exerting some pressure on the fertilizer industry. Several fertilizer companies have responded to this pressure by developing "natural" fertilizers. The Ringer Corporation, for example, manufactures some of its fertilizers from the organic by-products of other industries. Nitrogen is derived from feather meal from the poultry business, phosphorus from ground bone, and potassium from sunflower seed ash.[13] For the environment, this is clearly a step in the right direction, for at one fell swoop the consumption of fossil fuels is reduced, and waste products that might otherwise contribute to our solid waste problems are put to profitable use.

FIGURE 24. Hardware stores stock a wide array of chemicals designed to aid the homeowner with every imaginable lawn maintenance problem. Photo: G. Carleton Ray.

Pesticides

The world market for pesticides reached nearly $37 billion in 1997.[14] Sales of lawn care pesticides in the United States make up a surprisingly large part of the total world expenditure: about 32 percent. In 1997, more than $2 billion worth of pesticides were sold for use on American lawns.[15] To understand the marketing strategy of the chemical lawn care companies, one need only look at the bewildering array of pesticides present in any hardware store or garden center during the summer months (figure 24).

One of the principal marketing strategies of the pesticide industry focuses on the chemical destruction of pests, organisms that compete with, destroy, or devalue agricultural products, including lawn grass. Advertising often projects a rather neat world where technology simplifies the control of weeds, insects, and plant disease and produces an ordered world under human control.

In the 1970s many citizens began to fall out of love with synthetic chemicals, including poisons such as those used in many turf pesticides. Chemical pesticides used in lawn care are designed to destroy or control living organisms and may therefore pose a threat to many forms of life, including humans and their pets. Fewer than 1 percent of the estimated half a million species of plants, animals, and microbes in the United States are considered pests.[16] The other 99 percent carry out an array of essential functions, such as decomposing organic wastes, degrading pollutants, recycling nutrients, moderating the structure of the soil, preserving genetic diversity, and serving as vital parts of food chains.

Many of these organisms live in the soil. What is soil? From the point of view of an ecologist or agriculturist, soil is a complex substance, the product of many years of occupancy by plants, animals, and microbes. Soil is composed of minerals, sand, silt, and clay; it contains thousands of organisms whose physical and biochemical actions make soil a dynamic body. These organisms decompose organic matter, renewing the store of soil nutrients. Soil also retains water, and this capacity, together with the presence of nutrients, enables plants to take root and grow. Soil is usually only one to two feet deep, with most activity and nutrients concentrated in the upper six inches or so—the topsoil.

Beneficial organisms, including organisms in the soil, can become the inadvertent targets of pesticides applied to kill known pests. In the 1960s Rachel Carson voiced concern about unintended targets and human health, and since then a stream of investigators have echoed her apprehension. As a result, chemical companies have found their pesticide markets influenced by mounting public concern. In a survey conducted in 1989 by *Lawn Servicing* magazine, 64 percent of lawn care professionals said customers confronted them with questions about pesticides. Thirty-three percent had cancellations as a result of what they called "bad publicity."[17]

Irrigation

Grass needs water to grow (figure 25). The amount of water that a yard demands (or its owner thinks it demands) is subject to variables in addition to place and climate. The type of landscaping, the varieties of grass grown, the type of soil, the pressure from neighbors, and the degree of importance the owner places on having a green lawn all contribute to the amount of water applied.

In arid regions of the country, such as the Southwest, irrigation is absolutely essential for lawns: relatively large quantities of irrigation water are required to maintain a lawn under high rates of evaporation. Without irrigation, local vegetation adapted to frequent drought conditions would soon replace the lawn. In the more humid regions of the country, many lawns survive quite nicely without supplementary water, although they may become dormant and turn brown during periods of low rainfall.

FIGURE 25. The Industrial Lawn often requires watering to preserve an ever-green appearance.
Photo: © Karen Bussolini, 1992.

The lawn irrigation market is built on technology designed to supply water. Delivery systems range from sprinkler cans, through hoses with attached sprinklers, to elaborate underground sprinkler systems. Attachments include fairly sophisticated technology such as timers and soil-water-sensing elements that automatically turn sprinkler systems on and off. One thing that all these devices have in common is the need for a source of water. Two kinds of profit are thus involved in the irrigation market: profit from the sale of technology and profit from the sale of water.

The now-defunct National Xeriscape Council, an organization that promoted landscaping with a minimum application of water, estimated in 1990 that up to 30 percent of urban water on the East Coast is used for lawn irrigation. For the West, they estimated that 60 percent of urban water use is for lawn irrigation.[18]

Not all eastern lawns are heavily watered. The Turfgrass Council of North Carolina documented in 1987 that in that state lawn watering is low on most homeowners' list of priorities and that more than 75 percent of homeowners never water their lawns at all.[19]

As fresh water becomes an increasingly scarce and valuable resource, many municipalities are taking a harder look at regulating lower priority water usage. In southern California, water conservation is part of the water code, which instructs residents to use only recycled water on lawns. Even regions considered to have ample water supplies have had to regulate water use for lawns and gardens during recent dry periods. The eastern drought of 1999 led to mandatory water restrictions in Pennsylvania, Maryland, Delaware, and New Jersey. Neighbors eyed each other cautiously, wondering if anyone would notice infringements of the new laws. Telltale green lawns exposed those who ignored the bans. Eventually the rains came, but the drought had triggered an interesting twist on the standard peer pressure surrounding the ideal of the continuously green lawn.

Lawn Equipment

Mowing is a requirement for any lawn, but manicuring is an essential requirement of the Industrial Lawn. The lawn equipment industry thrives on sales of tools designed to achieve a perfectly groomed lawn (figure 26). A special tool exists for every conceivable lawn maintenance task: edgers, weed eaters, fertilizer spreaders, dusting and spraying equipment, leaf blowers, precision seeders, turf aerators, rotary and reel mowers, rider mowers, and on and on. In 1997, $7.4 billion worth of lawn and garden equipment was shipped to domestic and foreign markets by U.S. industry.[20]

The equipment market has traditionally depended on sales of new, improved state-of-the-art gadgetry. Currently the most popular state-of-the-art gadget is the large, expensive, and profitable rider mower. Rider mowers, which accounted for 25 percent of all mowers sold in 1990, now account for 30 percent of all mower sales.[21] The demand for rider mowers is only partly linked to new housing construction; profits can also be generated by consumers who "step up" from their old walk-behind power mowers to the "convenience and comfort" of new rider mowers.

Like the fertilizer and pesticide markets, the mower markets are feeling the impact of ecological thinking. A rash of new mowers designed to leave the grass clippings on the lawn has arrived at local hardware and garden centers. The environmental rationale is that the clippings fertilize the lawn as they decompose and release nutrients to the grass roots and that by leaving clippings on the lawn the stream of yard waste going to the local landfill is reduced. This is bad news for the fertilizer industries because it reduces the need for fertilizers.

FIGURE 26. Lawn care can involve simple hand tools, but today's industry has promoted a vast array of motorized labor-saving tools. Photo: © Karen Bussolini, 1992.

Sod

To install an instant lawn, homeowners can purchase sod, grass grown as an agricultural crop (figure 27). Sod can be used to establish replacement lawns, but the primary market for sod is new houses with unlandscaped yards. The price of sod is an important start-up cost associated with a new home for homeowners who want a lawn and want it now. As the housing market goes, so goes the market for sod. Housing starts have increased from about 1 million in 1991 to 1.6 million in 1999. This increase has been closely shadowed in the sod market.[22]

Sod is an expensive proposition. In 2000, do-it-yourself homeowners could purchase sod from garden supply stores in units of 9 square feet for $3.99. At this price, a 5,000-square-foot lawn would cost about $2,217, not including other such costs as delivery, site preparation, fertilizers, and so forth. Nevertheless, 4.6 million Americans bought sod in 1998.

A principal environmental question associated with sod farming is its effect on the agricultural soil that supports the sod farm. Because some soil is removed with each sod harvest, sod farming would appear to be a soil-mining process that would lead to the destruction of the original soil. At a time when there is great concern

FIGURE 27. The production of sod requires large inputs of water, fertilizer, pesticides, and fossil fuel during growth, harvesting, transportation, and replanting. Photos: Grant Heilman/Grant Heilman Photography, Inc.

about the erosion and destruction of agricultural soil worldwide, it might seem questionable to destroy agriculturally productive soils to produce instant lawns.

Fortunately, it is possible to make new soil! Most of the exposed earth we see around construction sites or other disturbed places is not soil as defined earlier but rather raw earth from which soil is made by ecological processes. The copious addition of organic matter can often turn these raw materials into reasonably decent soils. In our society today, there are two plentiful sources of organic matter: sewage sludge and leaves collected from street trees. With appropriate attention from government, industry, and individuals, these two major sources of solid waste could be put to profitable use transforming raw earth into fertile soil.

Labor

All grass has one consistent characteristic: it grows taller and taller. Keeping grass short is the lawn care industry's bread and butter. This industry was built on manual labor, from early teams of European scythers to today's summer jobs for American teenagers. In today's market, labor remains the largest single cost in both home and commercial lawn care.

Labor is what makes the lawn care industry an important addition to a local economy and might be used in an argument justifying intensive lawn care operations. As the main expense in the industry, however, labor is also the first area targeted for cutbacks in a competitive economy. The lawn situation might be somewhat analogous to the intense argument that has long raged in the Pacific Northwest about cutting old-growth forests. Industry maintains that reducing cuts will cost the loss of jobs, but at the same time industry is furiously eliminating jobs through the introduction of labor-saving technology.

HOMEOWNERS' EXPENDITURES

To determine how much the "average" homeowner spends is a daunting task. Several states (Maryland, 1987; New Jersey, 1983; New York, 1977; North Carolina, 1986; Ohio, 1989; and Pennsylvania, 1991) have conducted surveys, but due to different methodologies and assumptions it is hard to compile "average" numbers from their results. Nevertheless, some generalizations are still possible.

The bulk of turfgrass covers home lawns. In New Jersey, home lawns represented three-quarters of the total lawn acreage in the state and half the annual maintenance costs. Ohio homeowners accounted for nearly two-thirds of the state's lawn acreage, and they spent 64 cents out of every dollar spent on lawn care. About two-thirds of the lawn acreage in North Carolina is in home lawns, and homeowners accounted for one-half of the statewide expenditures.

Because the average size of the home lawn is only one-third of an acre, the expenses incurred in lawn care are quite high on a per acre basis, much higher than the per acre cost for agricultural crops, for example. The average household has a mower and a variety of other tools and makes annual purchases of new tools, fertilizers, and pesticides. To gain some idea of the magnitude of these expenses we will look at the North Carolina Turfgrass Survey. North Carolina is a state of numerous small cities with an extensive rural area and may not be representative of the nation as a whole. Nevertheless, the statistics from North Carolina can give us some idea of how we spend our money on lawn maintenance.

The average lawn owner and industry's ideal lawn owner are two different people. The industry ideal is a figure of lore, industry news releases, and do-it-yourself gardening guides. The North Carolina Survey, based on personal interviews, yields a more realistic picture. North Carolina's average home lawn is 0.6 acres, twice as large as the national average. Homeowners spent $407 on their lawns each year: $158 was for maintenance while the remaining $249 went to purchasing new equipment. The lawn is many things to many people, but even so it represents an annual expense of hundreds of dollars for most households and likely much more for higher-income homes.

As Americans cope with an economy that requires longer work hours and two wage earners, mowing the lawn can lose its attraction. Lawn care services provide an alternative that claims to cost no more than a homeowner's do-it-yourself investment of money and time, but with more professional results. Professional lawn care services weave all of the lawn industry components into "greenscapes"—the only thing left to the lawn owner is writing the check.

The lawn care service industry has experienced unprecedented growth, and demand is still increasing. In 1990, there were approximately 5,500 professional lawn care companies in the United States. These ranged in size from one person with a pickup truck to major public corporations with franchises across the country. In 1998, twenty-one million households spent $16.8 billion on professional lawn care service.[23]

The key to profit in the industry is to sell services, among them mowing, fertilizing, controlling insects and disease, aerating, seeding, dethatching, sodding, renovating, edging, and maintaining shrubs and small trees. Financial success rides or falls on persuasive salesmanship. Companies offer a variety of packages that differ most in frequency of chemical application and attention paid to surrounding plants. Often they offer a lawn owner a menu of lawn care more complicated than the dinner menu at the Waldorf Astoria. Not infrequently they have been the target of heavy criticism for careless or deceptive advertising about the efficacy and safety of pesticides used in the lawn care procedures they advocate. In 1990, the U.S. General Accounting Office (GAO) issued a report about the products they use. Staff from the GAO identified themselves as private citizens and called several lawn care companies to ask about product safety. Although some companies acknowledged that there were environmental and health risks, several others claimed that their products were safe or nontoxic. (Several health-related problems that contradict these safety claims are detailed in the next chapter.) Responses that the GAO found to be misleading or false included:[24]

"The only way to be affected by [the pesticide] 2,4-D would be to lay [*sic*] in it for a few days."

"The safety issue has been blown out of proportion. Such a small amount of chemicals are put down directly on plants. . . . [They do] not affect animals or people."

"All chemicals [used] are nontoxic."

"Dogs may get a rash or irritated [from diazinon], but they will only feel a little itchy. This is the same reaction the applicator gets when the pesticide touches their [*sic*] skin."

The lawn care industry thrives on and is probably the chief promoter of the

Industrial Lawn. The idea of the all-grass, pest-free, continuously green, and frequently mowed lawn is money in the bank.

THE CLASH OF PHILOSOPHIES

The Freedom Lawn and the Industrial Lawn strategies embrace extremely different philosophical views of nature. Although it is a product of human management, the Freedom Lawn has a large element of what Wes Jackson, a pioneer in the sustainable agriculture movement, has called "nature's wisdom." The pattern of the Freedom Lawn is in large measure the result of natural processes. Most of all, the Freedom Lawn emphasizes the use of solar energy and minimizes the use of fossil energy and other scarce natural resources.

The Industrial Lawn, by contrast, is under stricter human management. With the help of technology, humans try to control the pattern of biological relationships within the lawn. Plant diversity is minimized; grass predation by insects is controlled by insecticides; weed species are controlled by herbicides; new, genetically altered grass strains are designed to withstand applications of potent weed killers; fungal attacks on grass are thwarted by fungicides; the addition of fertilizers substitutes for naturally occurring nutrient cycling; droughts are avoided by irrigation; and mechanical soil aeration compensates for the absence of a soil structure that promotes natural aeration.

Relative to the Freedom Lawn, the Industrial Lawn depends upon fossil fuel energy, irrigation water, pesticides, and fertilizers. It requires a considerably greater drain on world resources. From an ecological point of view, the Industrial Lawn also causes unanticipated environmental side effects. These matters are the subject of the next chapter.

4 *Environmental Costs*

As you wheel out the lawn mower for the first time in the spring to subdue rampantly growing grass and defiant stands of dandelions, it is difficult to imagine that mowing your lawn might affect anything more than those dandelions, your peace of mind, and your neighbor, whose view from the dining room window includes your yard. Yet creating something as beautiful and whole-some-looking as a lawn can adversely affect the earth. The more we work to make the lawn beautiful by watering, spraying, and mowing, the more we remove it from the natural ecosystem that would exist if we had not interfered. What is the eco-logical price of making and maintaining a lawn?

It once seemed inconceivable that humans could affect the processes that main-tain life on earth. The earth appeared simply too vast. With the Industrial Revolution, the growth of technology, the enormous growth in our use of renewable and non-renewable resources, and the explosion in numbers of human beings on earth, we are finding, as William Vogt suggested in the 1950s, that the planet has shrunk, that space is no longer the protection we once thought it was, and that indeed we are capable of affecting global ecological processes.

Global ecological cycles are being altered in ways inimical to human welfare. Many regional air and water pollution problems have developed. We are concerned about the productivity of the lands and the seas. So grave are these problems that we are daily bombarded with news about global warming, the ozone hole, and the disappearance of tropical rain forests. Indeed, many are concerned about the future of humanity. Innumerable conferences, such as the Earth Summit in Rio de Janeiro in 1992, the climate change discussions in Kyoto in 1997, and the World Trade Organization meeting in Seattle in 1999, discuss the dimensions of, and the projected solutions to, an ever-lengthening list of environmental problems. This is the context within which we wish to consider the ecology of the lawn, our own little piece of the biosphere.

NATURAL ECOSYSTEMS VERSUS HUMAN-MODIFIED ECOSYSTEMS

The people, grass plants, earthworms, and other organisms that live in or pass through our lawns form a dynamic community or ecosystem where species interact with each other and with their physical and chemical surroundings. We sometimes forget that our lush green carpets can be viewed as such a system—that ecology is occurring in our own backyards. How does the ecology of the lawn compare to the ecology of more "natural" systems, and how do activities that take place in our yards affect the local, regional, and world environments in which we live? Our third of an acre may not seem like much, but lawns combine to cover more than 31 million acres of the United States, making grass the largest single "crop" and a significant part of the American landscape.

Although human civilization today affects even the most remote ecosystems ranging from the tropics to the Antarctic, we can identify ecosystems in which humans exert relatively little day-to-day influence. These "natural" ecosystems are powered almost exclusively by the sun, which provides the energy needed by all the plants and animals and drives the water cycle and the circulation of nutrients within the ecosystem. Natural ecosystems, like a forest or a prairie, do not require external management because the organisms are adapted to the physical and chemical conditions of the site and because the circulation of nutrients between the living and the nonliving is fairly complete. These ecosystems are solar powered, self-regulating, and self-fertilizing (figure 28).

The vast area of the United States contains a wide variety of climates and soils. In presettlement times, natural ecosystems in Maine and many of the far western states were dominated by evergreen forests, large sections of the eastern states were covered with deciduous forests, the middle of the continent was clothed in grasslands, and large areas of the Southwest supported desert ecosystems. Today, only remnants of undisturbed, naturally occurring ecosystems remain. Agriculture, grazing, forestry, mining, and other land uses, which are essential to maintaining our way of life, have substantially altered the ecosystems of presettlement times. Yet the same climate and soil conditions that gave rise to the forests, grasslands, and

-Solar energy
-Precipitation
-Nutrients in rain and dust
-Atmospheric carbon
 dioxide
-Seeds from local region

INPUTS

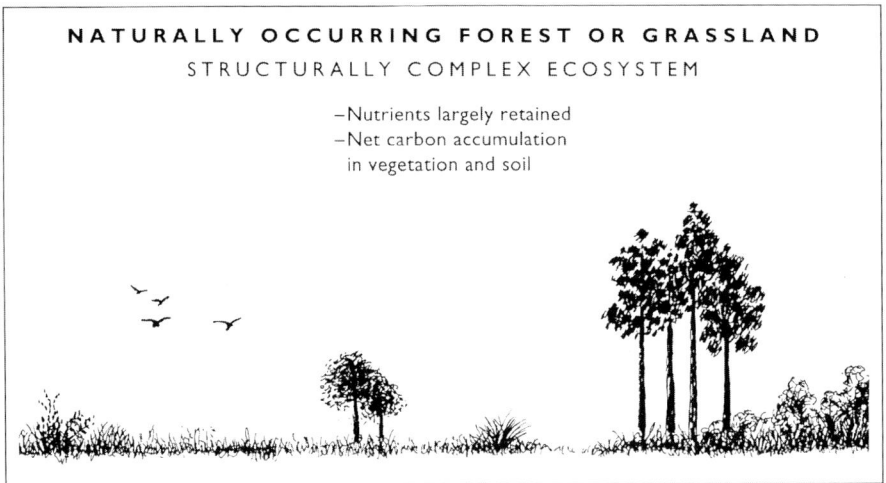

NATURALLY OCCURRING FOREST OR GRASSLAND
STRUCTURALLY COMPLEX ECOSYSTEM

-Nutrients largely retained
-Net carbon accumulation
 in vegetation and soil

-(Relatively) large
 proportion of water
 evaporated to atmosphere
-Little surface runoff

OUTPUTS

-(Relatively) few nutrients
 lost in drainage water
-Less carbon dioxide output
 than input
-Very little soil erosion

CONSEQUENCES

-Much biological
 diversity: provides home
 for many species of
 plants, animals, and
 microbes

-Clean water to streams
 and ground water
-Reduces global warming
 by net removal of
 carbon dioxide from the
 atmosphere

-No impact on municipal
 land fills
-No impact on global
 fossil fuel supplies

FIGURE 28. A general comparison between the relative environmental impacts of a naturally occurring forest or grassland ecosystem (above) and the Industrial Lawn (right) that might replace it. Asterisks indicate consequences that would be minimized (*) or absent (**) if a Freedom Lawn were to replace an Industrial Lawn. The Freedom Lawn is mowed, but the grass clippings are not removed, nor are fertilizers or pesticides applied.

INPUTS

- Solar energy
- Precipitation
- Nutrients in rain and dust
- Atmospheric carbon dioxide
- Seeds from local region

- Fossil fuel energy
- Irrigation water
- Nutrients in fertilizers
- Pesticides
- Grass seed or sod

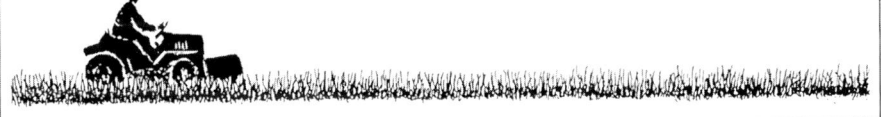

AN INDUSTRIAL LAWN
STRUCTURALLY SIMPLE ECOSYSTEM

- Relatively fewer nutrients retained
- Net carbon loss when carbon in fossil fuels, directly or indirectly associated with lawn care, is counted

OUTPUTS

- More surface runoff
- More nutrients lost in drainage water
- Pesticides and fertilizer nutrients washed into neighboring water supply

- Carbon dioxide output greater than input
- Nutrients and pesticides removed in grass clippings

CONSEQUENCES

- Less biological diversity: local plant species displaced by turf grasses and turf-adapted animals and microbes **
- Contributes to increased global warming

- Increases stress on municipal water supplies *
- Increases municipal solid waste problems *
- Pesticides may contaminate food chains *
- Pesticides on lawns may threaten human health *

- Disrupts biology of neighboring surface waters *
- Uses up global fossil fuel supplies

deserts persist today, and if the influence of humans were removed, ecological theory tells us that after a time the regional patterns of naturally occurring ecosystems would reappear. If, for example, we stopped maintaining our lawns in Tucson, Arizona, they would become minideserts, in Potomac, Maryland, they would start developing toward forests, and in Lincoln, Nebraska, the first stages of a prairie would appear.

Capability Brown's lawn was developed for the cool, moist, mild climate of England so favorable to growing grass. In North America, we maintain lawns in areas that were once forests, grasslands, and deserts and do so under a variety of climatic conditions, many of which are not naturally favorable to turfgrasses. Lawns grown in these regions differ dramatically from the natural ecosystems they replace because grass grown in an environment where it would not naturally occur depends on human management and on supplements of fossil energy, water, and chemicals for survival. The supplement required depends on the naturally occurring local climatic and soil conditions and on the kind of lawn we choose to maintain, with the Freedom Lawn at the low end and the Industrial Lawn at the high end in terms of supplements required. Yet in desert or semiarid climates, even the Freedom Lawn would not be possible without irrigation. In the desert, a yard would consist of native drought-resistant plants, stones, and patches of bare ground.

The use of fossil energy and chemical additives directly affects not only the ecosystem of the lawn but interconnected ecosystems, such as air and water, and these supplements may have both direct and indirect effects on the health of many organisms, including human beings. Using these supplements for the lawn may also limit the supplies of scarce resources that could be available for more vital uses.

FOSSIL ENERGY

Grass plants, through the process of photosynthesis, use the energy of the sun to grow. The lawn, particularly the Industrial Lawn, also requires the burning of carbon stored away in fossil fuels millions of years ago to power equipment and to manufacture and transport fertilizers and pesticides. The lawn is firmly linked to environmental issues that surround the consumption of fossil fuel (figure 29).

During the oil crises of the mid-1970s energy was thrust into the spotlight as a national issue. Dramatic steps were taken to reduce the consumption of fossil fuels, including improving mileage standards for automobiles and introducing tax breaks for energy-saving building improvements. As oil prices dropped, complacency about energy issues returned during the 1980s. In the mid-1990s the Persian Gulf crisis again put energy consumption back on the national agenda as the hazards of depending on the oil-rich but politically unstable Middle East again became apparent.

Political concern over fossil fuels has waxed and waned in response to supply, but scientific concern about the impact of the use of fossil fuel on our environment has steadily intensified. The use of fossil fuels has been increasingly linked to environmental problems including smog, acid rain, mega-oil spills, destruction of the ozone layer, and global warming. Energy problems do not just concern prices at the gas pump anymore, and environmental considerations such as global warming are not as far removed from decisions you make about the lawn as you might think.

In lawn management, the most obvious use of fossil fuel is in the mechanized equipment we use to groom our grass. Lawn mowers, aerators, leaf blowers, weed whackers, and edgers all consume fossil fuels, either directly (gasoline-powered equipment) or indirectly (electrically powered equipment). Approximately 18.6 million pieces of outdoor power equipment were sold in 1998 in the United States.[1] Most machines are powered by lightweight, two-cycle engines that are very inefficient. Although this represents only a small percentage of the gasoline we use in cars, for a variety of reasons, the amount is more significant than the percentage might indicate.

But in a proper accounting of energy costs, we should add the fossil fuels required to produce and ship fertilizers and pesticides. The principal nutrients in fertilizers are nitrogen (N), phosphorus (P), and potassium (K); these are listed on fertilizer bags as percentages by weight, for example, NPK 10-10-10.

Although the air we breathe contains 78 percent nitrogen gas, plants cannot use nitrogen in this form. Some bacteria, such as those associated with the roots of clovers or alder bushes, can convert atmospheric nitrogen into a form usable by plants. In natural ecosystems, microorganisms that decompose dead organic matter also release nitrogen previously incorporated in plant or animal tissue. Before World War I, most nitrogen fertilizers came from such organic sources as animal manure, guano, and bloodmeal. In 1913, Fritz Haber and Carl Bosch, two German scientists, learned how to capture nitrogen from the atmosphere by combining it with the hydrogen in natural gas to form synthetic ammonia, a nitrogen-rich compound. This discovery formed the basis for the synthetic fertilizer industry whose products require substantial fossil-fuel consumption and have increasingly replaced organic fertilizers.

Fossil fuels are needed to mine, refine, and transport potassium and phosphorus used in fertilizers. Each of these processes requires the input of fossil fuels. The crude mined product has to go through refinements powered by fossil fuel before it is placed in a bag. The bag of 10-10-10 you buy in Macon, Georgia, may have originated, in part, in Florida, Utah, and Saudi Arabia; thus still more fossil fuel is required to power the ships, trains, trucks, and finally the car that will move these products from their sources to your lawn. The manufacture and distribution of pesticides is similar to fertilizers in that fossil fuels are needed both as raw materials and in manufacturing and distributing these chemicals. Evaluating the application of fertilizers and pesticides to one's lawn thus involves not only the direct effects of these substances but their hidden cost in terms of fossil fuel consumption.

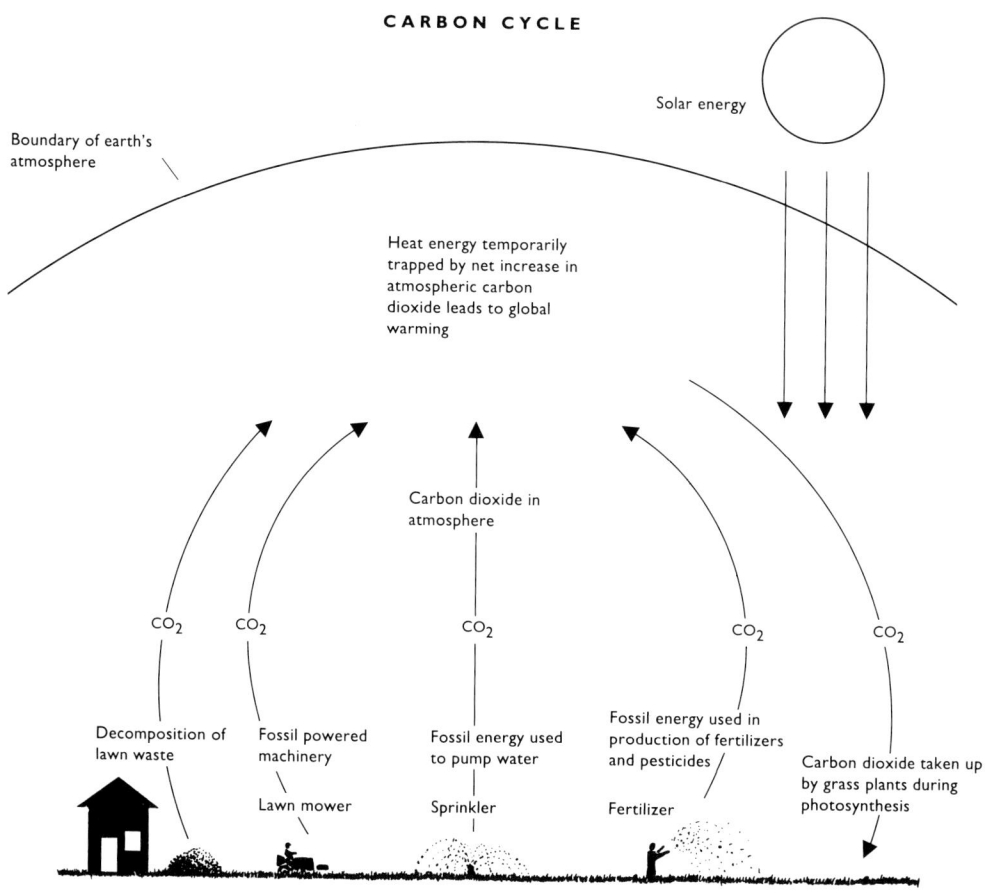

CARBON CYCLE

Solar energy

Boundary of earth's atmosphere

Heat energy temporarily trapped by net increase in atmospheric carbon dioxide leads to global warming

Carbon dioxide in atmosphere

CO_2 CO_2 CO_2 CO_2 CO_2

Decomposition of lawn waste

Fossil powered machinery

Fossil energy used to pump water

Fossil energy used in production of fertilizers and pesticides

Carbon dioxide taken up by grass plants during photosynthesis

Lawn mower   Sprinkler   Fertilizer

FIGURE 29. The carbon cycle as it pertains to the Industrial Lawn. As the diagram shows, the Industrial Lawn adds more carbon dioxide to the atmosphere than it removes, contributing to global warming.

As discussed previously, Turfgrass Producers International and the PLCAA have stated that through photosynthesis a fifty-by-fifty-foot lawn generates enough oxygen to meet the needs of a family of four. This kind of accounting fails to consider the oxygen consumed by microorganisms as they decompose grass clippings, nor does it account for the large amounts of oxygen consumed in burning fossil fuels associated with lawn mowing and the production and distribution of fertilizers and pesticides. When these factors are included in the calculations, the fifty-by-fifty-foot lawn becomes responsible not for a gain but for a substantial net loss of oxygen from the atmosphere.

WHAT IS GLOBAL WARMING?

Global warming results when the atmospheric content of greenhouse gases (carbon dioxide, methane, nitrogen oxides, and chlorofluorocarbons) is increased by human-generated pollution. These gases slow the loss of heat from the earth and warm the atmosphere (figure 29). Scientists are not absolutely certain that climatic change is actually occurring nor are they sure of how climatic change will affect the earth. However, they do know that "the average global temperature has already risen about 1 degree Fahrenheit in the last century, compared to a natural [change] of less than 1 degree per millennium over the last 12,000 years."[3] A degree or two may not seem like much, but it does not take a big change in temperature to cause big changes in our environment. During the last ice age, when much of the northern United States was covered with an ice sheet thousands of feet thick, the average global temperature was only 9 degrees Fahrenheit lower than today's. A warming of just 3 to 4 degrees over today's average temperature will make the earth warmer than it has been at any time in recorded human history. Many atmospheric scientists predict dire consequences from small average temperature changes, including sea-level rise and major changes in weather patterns. The ice caps would melt, agricultural patterns would shift, low-lying coastal areas around the world would be flooded, and major shifts in the earth's vegetational patterns would occur. Human societies, which depend on the biotic stability of the globe, would be severely stressed and thoroughly disrupted by these changes.

Although we do not often realize the energy consumption involved in watering our lawns, in arid climates such as the Southwest, irrigation can match, if not outstrip, the energy costs involved in lawn mowing. Water has to be pumped over various heights to move it from its source, such as the Rocky Mountains or an aquifer deep in the earth, to the place it is used in the dry Southwest, and pumping requires energy. The city of Irvine, California, has estimated that watering one acre of lawn for one year consumes as much energy as mowing that lawn.[2]

The use of fossil energy is essential to our survival and well-being, but fossil fuels have serious and costly environmental and health effects: the burning of fossil fuels has been shown to contribute to the formation of acid rain, smog, ozone, and greenhouse gases and to cause respiratory health problems for residents in many large cities.

Vehicles, power plants, factories, and space heating are primary causes of these problems, but the use of fossil fuels associated with lawn maintenance contributes to the problem. Indeed, the numbers are staggering. California's Air Resources

Board once determined that in the one hour it takes to mow a lawn, the power lawn mower emits pollutants equivalent to driving 350 miles. Annual pollution emissions from lawn utility machines in California are equivalent to the emissions produced by 3.5 million 1991 model automobiles driven 16,000 miles each.[4]

Why are the emissions from lawn equipment so high? Lawn equipment generally uses small, two-cycle engines. Although light and mechanically simple, two-cycle engines produce significantly more pollutants than the more efficient four-cycle engine or automobile engines.[5] Indirect pollution generated by electrically powered lawn tools must also be considered. Electric motors may not release pollutants in your yard, but the power plant that produces the electricity releases pollutants elsewhere that contribute to health and environmental problems.

Although the emissions from an individual homeowner's mowing may seem small, the collective emissions generate an appreciable amount of air pollution. Recognizing the significant amount of pollution emitted by lawn care equipment, California passed regulations in 1990 that required manufacturers to reduce emissions by the year 2000. Influenced by California's regulations and the results of their own emissions study, the U.S. Environmental Protection Agency (EPA) has developed small-engine emission standards for the entire United States.[6] The new rules, which apply to all mowers, took effect on September 1, 1997.

CHEMICALS

Although Rachel Carson opened Americans' eyes to the health hazards stemming from chemical use in the environment, pesticide use has doubled since Carson's *Silent Spring* was published in 1962 (figure 30). According to the Environmental Protection Agency, a typical annual management program for an Industrial Lawn includes four or more applications of a high-nitrogen fertilizer and ten or more doses of various pesticides. The EPA estimated that in 1984 more synthetic fertilizers were applied annually to American lawns than the entire country of India applied to all its food crops.[7] Most Industrial Lawns receive between three and twenty pounds of fertilizers and between five and ten pounds of pesticides every year.[8] The National Academy of Sciences found that homeowners use up to ten times more chemical pesticides per acre than do farmers.[9] In Connecticut, homeowners use 61 percent of the pesticides applied in the state.[10] The number of households purchasing yard pesticides increased from 19.7 million in 1997 to 24.7 million in 1998.[11]

Once applied to a lawn, fertilizers and pesticides can follow a variety of paths

FIGURE 30. Many homes contain collections of partially used boxes, bottles, and bags of chemicals. Photo © Karen Bussolini, 1992.

and have a variety of unanticipated environmental effects. Through the lawn's connection with the air stream and flowing water, chemicals can move into and affect distant ecosystems.

Fertilizers

Humans often find rapid growth desirable in the landscapes they manage, and lawns are no exception. The lush green of the Industrial Lawn is a sign of high growth rates. Growth is the rate at which solar energy, captured in photosynthesis, is stored in plants and is measured as a gain in weight. In naturally occurring ecosystems, plants seldom operate for long at maximum growth rate because some factor eventually limits production. Scientists have found that if the limiting factor is added to the ecosystem, high growth rates will resume. Thus, fertilization increases growth by removing the effect of limiting factors.

Naturally occurring ecosystems usually retain most of their nutrients within the system, but human-managed systems can lose substantial amounts. Many nutrients are lost through harvesting, be that fruits, vegetables, grains, logs, or the clippings

HUMAN HEALTH AND THE LAWN

Homeowners often watch, without concern, as workers from lawn care companies don pro-
tective suits before spraying pesticides on their lawns (figure 31). This response stems from
the belief that lawn chemicals are not dangerous or that the benefits of sprays simply out-
weigh the risk of spraying potentially health-threatening toxins in their yards. Unfortunately,
the people who choose to dismiss health concerns for themselves and their children do so
for their neighbors as well, for the environment affected by their application of pesticides
does not stop at their fence or driveway.

Although our lawns conjure up images of good health, lawn care chemicals are increasingly associated with human health problems. In 1990, the Senate conducted hearings on
the use and regulation of lawn care chemicals. Some of its findings were startling:

- Thirty-two out of thirty-four major lawn care pesticides have not been fully assessed
  for their long-term effects on human health and the environment.
- Of the people using pesticides, 50 percent do not read the warnings on the containers
  that are designed to protect human health.
- Commonly used lawn care chemicals have been implicated in a number of human health
  problems. Individuals testified that they experienced severe nervous system reactions,
  including nerve damage, after exposure to some chemicals. One man testified that
  his brother had died after playing golf on a course that had just been sprayed with
  a fungicide. There were other reports of acute toxic responses to lawn care chemicals
  as well as complaints of nausea, rashes, and headaches. Reports of pets dying after
  exposure were not uncommon.

Chemical manufacturers tend to dismiss reports of illnesses or death as isolated and
unsubstantiated cases, but many well-documented cases of pesticide poisonings exist. A
study of lawn care and tree service applicators in New York found 28 confirmed cases of
pesticide poisoning in the state between 1990 and 1993. In most cases, the workers did not
follow labeled safety precautions. In the application of pesticides, it would seem prudent to
err on the side of caution, if for no other reason than lawns are the playgrounds of America.

and leaves we remove from the lawn. Nutrients can also be lost through erosion, a
significant problem for the American farmer. The application of synthetic fertilizers replaces lost nutrients. To ensure maximum productivity, farmers and lawn owners often add more fertilizer than plants are capable of assimilating into their tissues. Plants, for example, can use only a part of the nitrogen present in fertilizer

FIGURE 31. Chemical control of weeds and pests is becoming increasingly popular among homeowners. Although chemicals serve many useful functions, most yard management procedures can be accomplished without them. Photo: Runk/Schoenberger/Grant Heilman Photography, Inc.

immediately; some will be incorporated into the soil, some may change form and be lost as a gas, and some may be lost in drainage water.[12]

The environmental impact of overfertilization can be seen within the lawn eco-system and in connected ecosystems such as streams, lakes, and estuaries as well as in the earth's atmosphere. Excess nutrients, especially nitrogen, have an array of negative effects on the ecosystem of the lawn. Too much nitrogen can increase the grass plant's vulnerability to disease, reduce its ability to withstand extreme tempera-tures and drought, and discourage microorganisms that are beneficial to lawn health. Some synthetic fertilizers may acidify the soil, limiting important biological and chemical processes.[13]

Fertilizer use can be linked to changes in the earth's atmosphere. When nitrogen fertilizers break down in the soil, the gas nitrous oxide can be released into the air. Experts have pointed out that nitrous oxide is a potent greenhouse gas that con-tributes to climate warming.[14] Nitrous oxide is also one of several gases that, upon reaching the uppermost levels of the earth's atmosphere, act to destroy the strato-spheric ozone layer that protects the surface of the earth from damaging ultraviolet radiation from the sun.[15] As we will see in a later section, nutrients from fertilizers can also pollute water.

Pesticides

Three groups of organisms may be thought to "threaten" the lawn: animals (such as moles and insects), weeds, and fungi. They may bring disease or change the ap-pearance of the lawn, disrupting the smooth, even carpet of green. Since the 1950s millions of homeowners have waged war on these enemies with pesticides, specifically with rodenticides, insecticides, herbicides, and fungicides. The EPA estimates that 74 million pounds of herbicides, insecticides, and fungicides were applied to resi-dential lawns and gardens in 1997.[16] Naturally occurring ecosystems protect them-selves against disease and insect outbreak in many ways. Some plants, like milkweed,

CHEMICAL DEPENDENCE IN YOUR OWN BACKYARD

Even a quick look at the daily paper is sufficient to make one aware that drug dependency is a major social problem in the United States. Addiction to cocaine, heroin, crack, and alcohol has left a trail of broken homes and people and a society totally confused about solutions.

PESTICIDE APPLICATION

THIS SIGN MUST REMAIN FOR 24 HOURS FOLLOWING PESTICIDE APPLICATION

Yet chemical dependency is not limited to drugs; a form of it may be present in your own backyard! The Industrial Lawn is in many respects a chemically dependent ecosystem. How does dependence develop? We could enter the cycle of dependence at any number of points, but let us start with the removal of grass clippings. Clippings are rich in nutrients, and if left on the lawn, they quickly decompose, making nutrients available to the next generation of grass plants. Removal, however, necessitates the use of fertilizers to replace the lost nutrients. Fertilizers often create environmental conditions in the soil that are inimical to decomposing organisms like earthworms and microbes; the weakening of these organisms leads to even less decomposition and the need for even more fertilizer. Excessive fertilization can cause the grass blades to grow at the expense of the roots, making the grass plant more susceptible to drought and necessitating irrigation to maintain growth and greenness. Pesticides are applied to control insects, fungi, and weeds, but their effects are often not limited to specific pests, and many nontargeted species may be damaged, weakened, or eliminated. Such nontargeted organisms may include those that carry out decomposition, the natural predators of insects destructive to grass and other desirable plants, and the fungi that contribute to plant growth by supplying nutrients. These chemicals weaken both the ecosystem's natural defenses against insects and disease and its naturally occurring nutrient cycling; to keep the altered system going, still more pesticides and fertilizers are required. Like any heroin addict, your lawn continually needs a fix to keep up its brave green front. But it is possible to break the cycle of dependency with all of its negative implications for our biosphere: just say no to chemicals.

produce chemicals in their leaves that make them unpalatable to organisms that might otherwise feed on them. Leaf-eating insects may be held in check by predators or disease, but in a larger sense, natural ecosystems often maintain good health through the large number of plants and animals that make up the system. If one plant or animal is weakened by insects or disease, another type of organism can take its place. Natural ecosystems usually do not put all their eggs in one basket.

Although a good, healthy lawn is the homeowner's goal, diversity in the lawn is often discouraged. A lawn composed of grass species alone is the desired goal in the Industrial Lawn. In this instance human goals are in direct opposition to nature's management scheme.

In their book *Lawn Care* (1988), Henry and Jane Decker describe potential environmental problems associated with continuous pesticide use.[17] These problems are outlined below and are illustrated in the box Chemical Dependence in Your Own Backyard (see page 78):

- Pest resistance: With the continued use of pesticides, resistant strains
  of target pests increase and the pest population becomes more difficult
  to keep in check.
- Inadvertent pest enhancement: When one pest is eliminated, another
  previously insignificant pest may attain a significant foothold.
- Killing beneficial organisms: Many insects and soil organisms, such as
  earthworms, bacteria, and fungi, carry out functions important to the health
  of the lawn. One fungus, for example, moves nutrients and water from
  the soil to the plant root. A fungicide applied to kill disease-causing fungi
  may also kill these beneficial fungi. As a result, the lawn owner may have
  to supply more water and fertilizer to achieve the desired productivity.
  Natural predators, such as spiders, help control harmful pests, but they
  may become the unintended targets of pesticides. Ironically, pesticides
  may kill off one of the lawn manager's greatest allies, microorganisms
  that decompose thatch. Clearly, pesticides interfere with the lawn's natural
  means of maintaining good health.
- Pesticide persistence: Many pesticides persist in the environment for a
  long time with their lethal capabilities intact. Newer pesticides have a much
  shorter period of potency, but they can nevertheless affect ecosystems
  beyond the lawn where they were applied. Pesticides may blow off-site or
  be leached from the lawn in drainage water and end up in wells or in streams
  and lakes where fish and other aquatic species may be affected. Pesticides
  are known to kill shellfish and other species in marine environments.

Some of the synthetic organic chemicals used as pesticides are fat-soluble and degrade slowly. As these pesticides move through the food chain, they become more and more concentrated and may even reach toxic levels in animals of prey. DDT, a pesticide used extensively in the 1950s and 1960s, washed into aquatic environments, where it was taken up by small algal plants. These were consumed by small fish that were in turn eaten by larger fish and eventually by birds of prey far removed from the original site of pesticide application. The buildup of DDT in their fatty tissues caused birds like the osprey to produce thin-shelled eggs that cracked easily, causing reproductive failure and an alarming drop in their population. The banning of DDT in the United States in 1972 has allowed some bird populations to recover, but other chemicals that are hazardous to wildlife continue to be used. One of the most widely used insecticides, Diazinon, kills waterfowl and other bird species. The EPA banned the use of Diazinon on golf courses and sod farms after receiving reports of bird deaths. But, one may fairly ask, how do birds distinguish a Diazinon-free putting green from a large suburban front lawn that has just been treated with Diazinon?

WATER SUPPLIES

"We talk [water] scarcity, yet we have set [some of] our largest cities in the deserts, and then have insisted on surrounding ourselves with Kentucky bluegrass. Our words are those of the Sahara Desert; our policies are those of the Amazon River."—Richard Lamm, governor of Colorado, 1975–87

Earlier we touched on the energy costs involved in irrigation and in transporting water from place to place, but there is still another problem with watering lawns. We are a nation whose water needs are rapidly rising while available supplies are shrinking and where regional water crises are becoming increasingly frequent. Population increases in the United States have combined with increased per capita consumption of water to generate a water crisis. In 1985 there were 61 percent more Americans than in 1950, yet during the same period, public water use (which excludes agricultural and self-supplied industrial use) rose 164 percent—a rate more than twice that of our population increase.[18] Between 1920 and 1960, reservoir capacity grew about 80 percent per decade. Dam building no longer provides the easy solution to the water problem that it once did: not only has the number of potential dam sites decreased, but the many negative effects of dam building on the environment have also become better known.

Water tables are falling and streamflow is decreasing in many overused river basins, especially in the arid West. In the Colorado and Snake Rivers, as well as in many other rivers, the decreasing water flow damages aquatic ecosystems, affecting the habitat for fish and wildlife. Such depletion also results in a less secure future for human residents of depleted areas. In the semiarid Plains states, water from the vast Ogallala Aquifer, which stretches from South Dakota to northern Texas, is being extracted much faster than it can be replaced by natural processes. Between 1950 and 1990, about 120 cubic miles, or 1,300 trillion gallons, were withdrawn. Water tables have fallen in the Dallas–Fort Worth area and in the Tucson metropolitan area, which is almost completely dependent on groundwater for its water supply.[19] Long droughts in California have lowered or emptied reservoirs, sometimes resulting in the cutoff of irrigation water to farmers and strict rationing plans in southern California's cities.

Water supply is not just a western problem, however. Water shortages are also frequent on the East Coast. In the New York metropolitan area there is growing concern that the increasing demand for water coupled with few remaining prime water-yielding sites may require the use of the contaminated Hudson River as a source of drinking water. The necessary treatment facilities would cost billions of dollars.[20]

A surprising amount of water for residential use goes to watering lawns. This is especially true in drier regions (figure 32). Natural water balance in a lawn is determined by the amount of rainfall received and the amount lost by evaporation and runoff. In arid regions of the Southwest, the water deficit is so great that naturally occurring vegetation is dominated by desert plants that have developed a tolerance for the harsh desert conditions. Growing a lawn under such adverse conditions requires virtually constant watering. Indeed, in the West, lawn watering can account for up to 60 percent of urban water use.

Lawn watering is not limited to arid regions, however. Casual observation in southern New England, where the climate is reasonably moist in summer, suggests that irrigation is used not only to avoid the effects of extended dry periods but also to heighten productivity; thus the sprinkler may be turned on even during periods of light rain! Watering lawns that we have purposely designed to be thirsty is certainly a practice that needs to be reevaluated.

WATER POLLUTION

When contaminated with chemicals and sediments, water becomes less usable for people and may be destructive to interconnected aquatic ecosystems. Pesticides

FIGURE 32. An Industrial Lawn in the semiarid climate of Salt Lake City, Utah. Photo: Anna Vernegaard.

and fertilizers used on lawns can contribute to this problem. They can travel by surface runoff or seepage through the soil to drinking water wells and other public water supplies, wetlands, streams, rivers, and lakes and even to marine environments.

Several factors determine the amount and impact of fertilizers and pesticides lost from lawns: soil texture, soil porosity, climate, watering practices, adjacent land uses, proximity to vulnerable aquatic ecosystems, and the kind and quantity of fertilizer and pesticide.

Sandy or gravelly soils allow water and dissolved fertilizers and pesticides to move relatively easily from the soil surface into groundwater. This movement is of particular concern in areas with permeable sandy soils lying over groundwater aquifers that supply drinking water, for example, in Cape Cod, Massachusetts, other parts of southern New England, and Long Island, New York. Millions who live in these areas depend on the aquifers beneath them for clean drinking water. Concentrations of nitrate in groundwater on Long Island increased dramatically in the 1980s and 1990s from lawn and garden fertilization; estimates are that 60 percent of the nitrogen applied ended up in the groundwater.[21]

Chemical contamination is a notable problem in drinking water wells. Nitrate, a form of nitrogen, is the most common contaminant. EPA surveys of groundwater wells used for drinking water in the United States indicated that 1.2 percent of community water-system wells and 2.4 percent of rural domestic wells nationwide contained concentrations of nitrate that exceed public health standards for drinking water.[22] High concentrations of nitrate in drinking water may cause birth defects, cancer, nervous system impairments, and "blue baby syndrome," in which the oxygen content in the infant's blood falls to dangerous levels.[23] Nitrate contamination may result from a number of land uses, including agricultural fertilization with nitrogen fertilizers, residential septic tanks, animal wastes such as those from cattle feed lots and hog farms, and lawn fertilization.[24] Once the contaminant reaches groundwater and is detected in a well, it is often difficult to determine its source. In many areas of the nation, agricultural practices are responsible for most of the nitrate that reaches groundwater and wells.[25]

Different types of nitrogen fertilizers decompose to release water-soluble nitrogen at varying rates. If soluble nitrogen is released at rates faster than grass plants can take it up, the excess may find its way into the groundwater. In general, organic nitrogen fertilizers, such as urea, release nutrients more slowly than inorganic fertilizers, such as ammonia and nitrate.

Adjacent land uses also influence the relative importance of fertilizer losses from a lawn. In residential areas with many septic tanks or in areas with intensive agriculture or other sources of nitrate pollution, lawn fertilization may account for a relatively minor part of a groundwater pollution problem. On the other hand, by reducing or eliminating the use of lawn fertilizer, the homeowner can minimize the lawn's contribution to nitrate pollution as well as save some money.

When fertilizer reaches interconnected bodies of water such as streams, lakes, and estuaries, damage can result. The addition of certain nutrients, such as nitrogen and phosphorus, to bodies of water can cause excessive growth of water plants in a process called eutrophication. Initially aquatic plant life flourishes because of fertilization. When these abundant plants die and sink to the bottom, they decompose, using up the oxygen in the water. Fish and other aquatic animals need oxygen, and under the oxygen-poor conditions that result from eutrophication, many die. Ironically, the result of too much fertilization can be a smelly body of water that is deprived of oxygen and of many forms of life.

The Mississippi River drains the heartland of the United States and like most large rivers has experienced massive changes, including eutrophication, during the past century.[26] Nitrate concentrations in the lower Mississippi doubled between 1950 and 1990, and agricultural fertilization has played a large role in this. About 44 percent and 28 percent, respectively, of the nitrogen and phosphorus fertilizer applied in the Mississippi watershed ends up in the Gulf of Mexico. Predictably, waters low in dissolved oxygen are found in the bottom ocean waters at the mouth of

FIGURE 33. In the course of a year, Industrial Lawn owners throw away grass clippings that contain much of the fertilizer they paid for. Photo: © Karen Bussolini, 1992.

the Mississippi River. Although the Mississippi example does not implicate the lawn, it does emphasize the connection between fertilization and eutrophication. Many scientists believe that a significant reduction in the eutrophication of water bodies in the United States "is not likely to occur without a reduction in fertilizer use."[27]

Pesticide contamination of groundwater is less well documented than fertilizer pollution but is of growing concern. Detectable levels of pesticides or pesticide breakdown products have been found in a small percentage of the wells in community water systems. A very few have one or more pesticides above health advisory levels. DCPA, an herbicide extensively used on home lawns, has been detected in EPA well-water surveys.[28] Because of their chemical character and their ability to travel through the soil, today's pesticides are more likely to contaminate groundwater sources than older pesticides, which leached through the soil more slowly.[29] For example, the EPA considers the commonly used herbicide 2,4D to be a "priority leacher" that travels quickly to groundwater.[30] A component of Agent Orange, a defoliant used in the Vietnam War, 2,4D has been linked to cancer and birth defects.

Although many may argue the significance of this pollution, most people agree that chemical pollution of water supplies and water bodies should be avoided. Some proportion of the fertilizers and pesticides used on lawns gets into water supplies and water bodies through runoff or leaching into groundwater. Whatever the percentages may be, it makes sense to do whatever we can to minimize this pollution. Having the greenest lawn on the block is certainly not worth contaminating our drinking water.

SOLID WASTE

We generate more than three-quarters of a ton of solid waste per person each year, 220 million tons of municipal solid waste in 1998. Most of this astonishing quantity of waste ends up in landfills and incinerators. Yard waste is the second largest component of the waste stream; three-quarters of yard waste is grass clippings from our lawns.[31] Clippings generally get bagged in large polyurethane or paper bags and find their way to the curb on garbage day (figure 33). Landfilling was originally viewed as an environmentally sound alternative to burning, but many landfills are now filled beyond the designed capacity. It has been estimated that by 2010, 80 percent of legal landfills will run out of room.[32]

There is no reason for grass clippings to be considered waste. Clippings can be left on the lawn where microorganisms will decompose them, releasing nutrients that have been stored in the cut grass so that they can be used again.

Grass clippings are a source of nitrogen and other nutrients. Removal of clippings may result in a loss of up to 100 pounds of nitrogen per acre of lawn per year. An additional problem is that decomposing lawn wastes in landfills can produce methane, one of the most powerful greenhouse gases.

Other wastes associated with lawn care also constitute a serious problem. Empty or partially empty containers of insecticides and herbicides are hazardous wastes.[33] Pesticide-laden grass clippings are another significant addition to the hazardous waste stream. In addition to constituting a danger in their own right, these toxins may interfere with the process of biological decomposition that would otherwise help to break down our garbage, thereby adding to the rapidity with which we fill up our solid waste dumps.

SPECIES DIVERSITY

The coming and going of species has been a normal development throughout the history of the earth. Species extinctions can be caused by large catastrophes or by more gradual environmental changes. There is strong scientific evidence, for example, that the dinosaurs became extinct after a meteor collided with the earth. Dust and debris from the collision clouded the planet's surface, reducing the amount of sunlight available to plants. Photosynthesis decreased, reducing the amount of available food for all organisms, and the temperature at the earth's surface declined precipitously. Fossil records indicate that "natural" extinction occurs at the rate of about one species per thousand years. In turn, new species appear at a similar, slow rate.

This balance between species extinction and species creation has recently been shattered by human activity. Most scientists agree that because of human activities such as habitat destruction and pollution, the extinction rate now greatly outstrips speciation rates. Although our estimates are crude at best, even conservative estimates of current species loss are startling. In 1991, scientists Paul Ehrlich and Edward Wilson estimated that "one quarter or more of the species or organisms on Earth could be eliminated within 50 years."[34] This is mass extinction, a biological holocaust of vast proportion, and a matter of extraordinary concern to biologists everywhere.

Citizens from every walk of life have expressed grave concerns about species extinction on our planet. Discussion often focuses on tropical rain forests and how their destruction feeds the high rates of extinction. Although it is true that tropical forests are among the most species-rich and most threatened ecosystems, extinctions are not unique to the tropics, and the events that lead to extinction are not always as obvious as the burning of forests. Indeed, as concerned citizens who wish to encourage species diversity, we may be ignoring the best place to begin this endeavor—our own backyards.

As we create and manipulate our lawns, we affect the organisms living within them. A new housing development frequently involves the wholesale destruction of a forest, an abandoned agricultural field, or a desert. The lawn, particularly the Industrial Lawn, is a highly simplified ecosystem compared to the forest or field, which typically contains a diverse mosaic of plants and animals. The creation of a lawn is thus synonymous with a reduction in species diversity (figure 34).

Animal populations are often determined by the vegetation available for food and shelter. The lawn provides little in the way of food for many species. We may see raccoons, squirrels, and other small mammals traveling across our lawns, but to survive they need areas more protected than lawns for denning sites.[35]

Birds are perhaps the most entertaining and welcome forms of wildlife that inhabit and visit residential areas. Changes in bird populations provide one of the best examples of the effect of converting natural ecosystems to suburban housing and lawn ecosystems. Many studies have shown that the bird population is greater in developed, suburban areas than in undeveloped "natural" areas.[36] For example, in

FIGURE 34. Across the country, lawns have replaced native habitats like woodlands and wetlands. Animals such as this forest-dwelling wood frog cannot survive in lawn ecosystems. Photo: Steve Zack.

a residential area in Tucson, Arizona, bird densities were observed to be twenty-six times that of surrounding desert covered with creosote bush.[37] Yet as anyone who has marveled at the numbers of seagulls at the local dump or pigeons in a park can tell you, numbers of individual birds and numbers of bird species are two different things. The flat, uncomplicated structure of the lawn attracts ground-feeding bird species that eat seeds and insects. House sparrows, starlings, rock doves, and, in the Southwest, Inca doves are among the most abundant species in urban and suburban areas.[38] Indeed, house sparrows, a species introduced from Europe, are often the first to colonize new residential areas. The suburban lawn displaces numerous diverse habitats that would have characterized the site before development: habitats that contained a variety of nesting sites, food, and shelter from predators. These displaced habitats contain fewer birds than developed sites but usually many more bird species.[39]

The more native vegetation is replaced by lawn, the less habitat is available for specialist species that rely on a few native plants for habitat or foods. Around Tucson,

Arizona, the desert vegetation is characterized by palo verde and saguaro cactus. As long as native vegetation remained substantial, moderate suburban development resulted in few changes in bird populations. When native vegetation was greatly reduced and replaced by Industrial Lawns, however, native birds such as the black-tailed gnatcatcher, pyrrhuloxia, brown towhee, and black-throated sparrow were entirely replaced by house sparrows and Inca doves, birds virtually absent from native vegetation. Starlings and house sparrows ventured into surrounding native vegetation and competed with native woodpeckers for nesting sites in saguaro cactus.[40] Similar results were observed in another Tucson development located in creosote bush desert. The wide-ranging loggerhead shrike and brown-headed cowbird were absent from an urban bird census, whereas the Inca dove, house sparrow, and starling were present in abundance.[41]

It may be time to link the local extinction that occurs in our backyards with the world decline in biotic diversity. The spread of the lawn and its accompanying destruction of native habitat may have a serious cumulative effect on the nation's flora and fauna, especially on plant and animal populations of limited size that occur in small and specialized habitats. The widespread expansion of suburbia and the American lawn has ousted local bird species and allowed a few species to dominate that are particularly suited to living among humans and houses. In a comparison study in Illinois, it was noted that bird species in urban areas were constant from region to region while in all other habitats there were great regional differences. The loss of diverse vegetation in urban areas accompanied the loss of bird diversity.

If the Industrial Lawn continues to accompany the development of the nation, we can predict two results. Not only will the important environmental functions of the undeveloped lands—the control of water and nutrient cycles and of energy flow in ways beneficial to society—be greatly diminished, but more and more species of plants and animals will be restricted to smaller and smaller areas, and larger and larger groups of fewer species will dominate the landscape. Already 20 million acres of residential lawns have combined with millions more acres of lawns in public parks and highway margins to displace a vast amount of native habitat for a large number and variety of plants and animals.

THE ENVIRONMENT AND THE AMERICAN LAWN

The American lawn has a long and noble history, and it has won a firm place in our hearts, but there is a darker side to this grassy swath. In our efforts to make it greener, to make it all grass, to keep it closely mowed, and to make it a constant companion of suburban development, we are unnecessarily contributing to some of the most severe environmental problems facing the world today.

The use of fossil fuels both directly in gasoline-powered equipment and indirectly through irrigation and the production and transport of fertilizers and pesticides contributes to regional air pollution and global warming. Excess fertilizers and pesticides wash off our lawns and run into our wells, streams, and lakes, where they may contribute to major environmental and human health problems. Lawn irrigation can exacerbate already severe regional water-supply problems, and lawn wastes are major contributors to our increasingly severe national solid waste problem. Finally, the replacement of millions of acres of naturally occurring ecosystems by the American lawn during the development process plays an important role in the continuing decline of local and regional species of plants and wildlife.

This does not mean that all aspects of the lawn are negative or that the eradication of lawns would solve any of the environmental problems to which they contribute. What all this information does mean is that we are now armed with the knowledge of how best to maintain or alter our lawns with the goal of creating a healthier, more diverse environment. Knowing the ecological implications of our actions, however great or small, enables us to act.

5

A New American Lawn

We started this book with a history of the lawn and the story of its eighteenth-century rise to prominence as the centerpiece of a particular view of nature. We have brought the story into the twentieth century by showing how industrialization transformed this favored piece of green in relation to the environment. Now we shall bring the lawn's aesthetic history up to the present.

The changes that brought about specialized seed varieties, fertilizers, and machines on a large scale affected not only agriculture and landscape. They also informed and transformed other areas, particularly the arts. Architecture, painting, and sculpture have all wrestled with their relationship to the industrial world and machines, which began to produce synthetic materials and chemically produced colors. Out of this struggle came the new forms that we call the modern movement. Landscape as art was the latest star to arrive in the constellation of the modern movement, but it lagged behind the others and saw itself more as a craft. Residential yards and gardens particularly remained as craftwork, with one exception: the lawn. The eighteenth century's favored piece of green received all the attention of industry: seed companies developed a lawn based on very few species; sod farms made the lawn a shippable, instant landscape. The lawn became, in fact, the modern landscape. And its importance went well beyond a residential application, though it has been the piece of modern landscape attached to most residential sites; it has become the favored landscape piece for all modern buildings, whether corporate headquarters or museums. The eighteenth-century lawn has taken on another meaning for the twenty-first century: it has become a visual symbol of the control of living things by humans, a perfect, completely malleable piece of nature. Or so we thought.

With our growing concern about the environmental health of the planet and our emerging understanding of ecological damage that may be caused by the American lawn, some have begun to ask, Can I manage or redesign my lawn in ways that minimize its negative impact on the environment and yet create an environment that meets my aesthetic and recreational needs? Or, in the broader context, Can I be a good steward of that small piece of the biosphere entrusted to my care?

We will now weave together our new ecological understanding with concepts of design and management. Presenting ecological reasons and data might be considered sufficient to bring about redesign of our lawns, but ignoring landscape design principles perfected over time to fulfill ideals of beauty denies the complexity of human behavior and dooms much of ecology's hard work. If lawns cannot be replaced with landscapes of beauty and usefulness, few of us will want to change. Yet landscape designers can no longer ignore the sound questioning of current landscapes and their management. Ideas about ecology do not design a landscape; ideas about beautiful landscapes as currently practiced have often had difficulty incorporating ecological ideas. Here we bring together some recommendations from ecology and art to give all of us the tools and concepts to guide us in reshaping our gardens and yards. The integration will be something that will take time and thought beyond these pages.

We suggest several alternatives to the classic American lawn: the all-grass, free-of-pests, continuously green, frequently mowed Industrial Lawn. It is not our mission to detail every possible way of transforming a lawn into an ecologically benign and aesthetically beautiful property. Our goal is to provoke new thinking about the lawn and its connection to the larger environment, to provide alternative strategies for managing or changing our lawns with the objective of implementing a new understanding of our relationship to the world around us, and to propose a new aesthetic approach to its design. No formulas or generic solutions are offered, only brief glimpses of what might be.

Many people, organizations, and books can provide assistance and suggestions concerning ways to change or adapt the lawn:

- Government agencies, including town conservation commissions, state cooperative extension services, or the U.S. Department of Agriculture.
- Gardening or landscape and nature sections of local newspapers, libraries, and bookstores.
- Nongovernmental organizations that focus on different aspects of lawn care.

Alternatives cover a range of options from changing the way you care for your grass to replacing the lawn completely. Four underlying principles, however, unite all the alternatives we propose: (1) meet the aesthetic, environmental, and economic needs and wishes of the individual homeowner, (2) to the degree possible, shift from fossil energy to solar energy, (3) reduce the use of chemicals and irrigation water, and (4) where possible, increase biological diversity. These principles reflect a shift in priorities away from a desire for a perfectly manicured expanse of lawn toward a healthy landscape in greater harmony with nature.

CONSIDERATIONS

For most Americans, the yard and the lawn are a part of their life as well as a considerable investment in both time and money. Before launching on a course of action, it is important to assess what you want to gain by changing the management and design of your lawn. A primary goal would be to lessen your contribution to the deterioration of the environment, but your yard and lawn serve other personal and family needs and may play an important role in your overall life-style and psychological health. You could strive to make your yard as ecologically sound as the forest primeval, but if it does not satisfy your expectations of beauty and practicality, it is not likely to succeed. So, too, economic feasibility must be considered. Fortunately, there are management and design alternatives that simultaneously meet human desires for beauty and a healthy environment and may save money.

The first step in considering a change is to assess your needs. Lawns and yards serve many purposes: an outdoor canvas for gardeners, a safe playground for kids and pets, a pleasant park for entertaining friends and family. You may, for example, want a lawn as a playing field for touch football or badminton, but there may be areas of little use where ground covers other than grass could be substituted.

As a start, it is a good idea to survey what you have. Figure 35 illustrates a typical Industrial Lawn design and management scheme.

Before changing your Industrial Lawn, identify how various subsections of the yard are used. Consider recreational uses, views large and small, and settings for flowers, shrubs, and vegetable gardens. Give some thought to ecological conditions: sun and shade, wet or dry spots, shallow or deep soils. You can learn a great deal simply by walking around and examining the yard. Most important, consider the climate and think about your area's native vegetation (see figure 17). All of these steps can help you evaluate your existing lawn and landscape design and help you to design alternatives that will fit your situation.

You may wish to consider neighborhood standards in your planning. Peer pressure

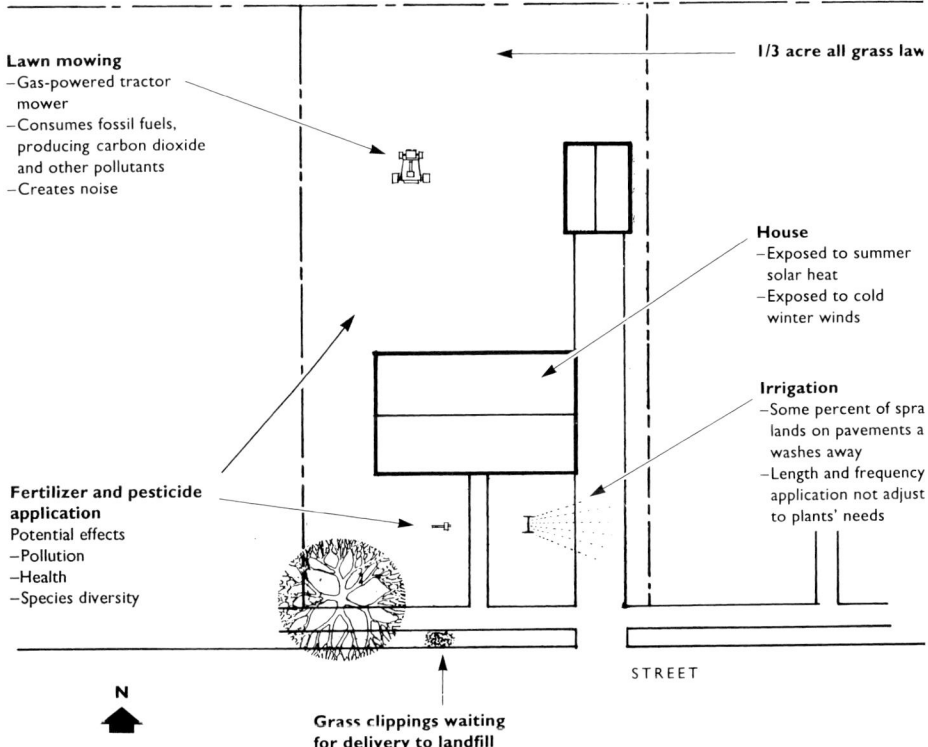

INDUSTRIAL LAWN

Property line

Lawn mowing
- Gas-powered tractor mower
- Consumes fossil fuels, producing carbon dioxide and other pollutants
- Creates noise

1/3 acre all grass law

House
- Exposed to summer solar heat
- Exposed to cold winter winds

Irrigation
- Some percent of spra lands on pavements a washes away
- Length and frequency application not adjust to plants' needs

Fertilizer and pesticide application
Potential effects
- Pollution
- Health
- Species diversity

STREET

N

Grass clippings waiting for delivery to landfill

FIGURE 35. The Industrial Lawn: a typical lot with house, front yard, backyard, and driveway. The notes highlight some of the topics raised by this book. See figures 36 and 39 for examples of how this yard's design and management might be changed to promote a healthier environment.

and even legal regulations might affect your decisions. If you wish, you can use your more private backyard to explore very personal and dramatic changes while making less radical but nonetheless significant changes in the front yard. You might find neighbors who will want to join you in making their yards safer for children and in increasing the populations of songbirds by reducing or eliminating the use of pesticides and by increasing the variety of shrubs and other plants attractive to birds.

Finally, there are the inevitable budget considerations. If you make extensive changes to your yard, your initial costs may be relatively high, but in the long term, you should realize savings from lower energy, fertilizer, pesticide, and water costs. Even if you choose to retain a lawn, you can immediately save money by initiating some simple management changes, such as decreasing how often you mow, water, and fertilize. For reducing your impact on the environment you will receive no tax rebates or income tax credits, but you will have the satisfaction of knowing that you are contributing to the creation of a richer, safer, and more diverse environment for all our children.

ALTERNATIVES

Landscape designers often speak about a vision for a piece of land: patterns created by vegetation, buildings, and open space, desired uses for the property, and how these aspects fit together as a whole. Landscape design, it must be remembered, is an art. This is reminiscent of Capability Brown's view of design. He approached a landscape by looking at its capabilities: the functional and aesthetic potential of the property. What vision do you have for your yard?

- You might want to keep the amount of lawn you have but change your management procedures.
- Another option would be to reduce the amount of lawn and use the surrendered space for other plants or for alternate nonliving materials.
- Finally, you might want to replace the entire lawn with other types of vegetation and landscaping materials. If you are building a new home, you have the option of preserving all or part of the existing natural vegetation.

All options involve both aesthetic and ecological dimensions. The aesthetic dimension is not formulaic. Each age has its ideal of beauty, and artists, as well as the rest of us, interpret the aesthetic ideas of an age to create an image. We point to models other than the eighteenth-century English lawn ideal and suggest that you develop your own ideas or, armed with your own specific needs, preferences, and aesthetic leanings, that you work with somebody who is interested in the artistic challenge of shaping a new Freedom Lawn.

Changing Lawn Care Practice

A primary goal in changing management is to make the lawn more dependent on solar energy and on the site's natural growing conditions, such as rainfall and soil, and less dependent on fossil energy and applied chemicals and water. Changes might involve selecting the right grass; reducing energy, fertilizer, pesticide, herbicide, and water inputs; or selecting an ecologically minded lawn care company. A simple

IMPROVED MANAGEMENT

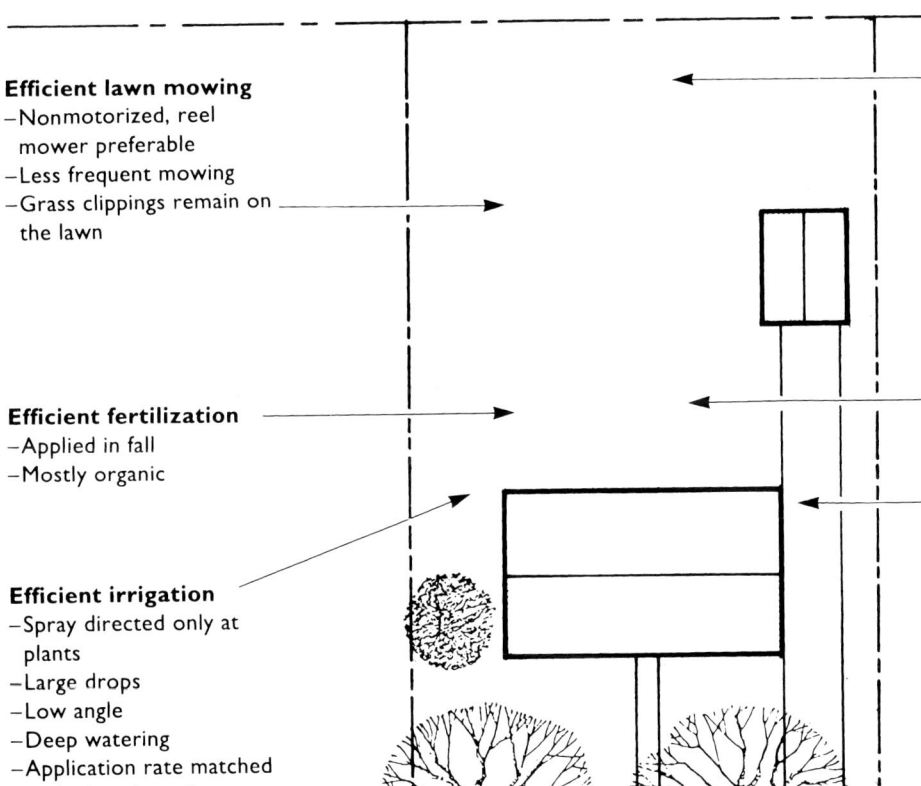

Property line

Efficient lawn mowing
—Nonmotorized, reel
  mower preferable
—Less frequent mowing
—Grass clippings remain on
  the lawn

Efficient fertilization
—Applied in fall
—Mostly organic

Efficient irrigation
—Spray directed only at
  plants
—Large drops
—Low angle
—Deep watering
—Application rate matched

FIGURE 36. Improved lawn management. The plan illustrates practices that do not change the yard's appearance but lessen its impact on local and global ecosystems.

management change, which would achieve the above goals, would be to switch to a Freedom Lawn. In figure 36 we show you how changes in your lawn management practices can reduce the negative environmental impact of lawn maintenance.

SELECTING THE RIGHT GRASS

If you want to keep an all-grass or nearly all-grass lawn, you might want to consider just what grasses you want. Lawn grasses include many species and many varieties within species. It is a good idea to choose varieties appropriate to your local environment and the special circumstances of your lawn. For example, you should consider local water availability. Some grasses require large amounts of water and should be avoided in areas frequently subject to drought. Some cities, such as Au-

New American Lawn { 95 }

rora, Colorado, mandate the use of specific grasses to avoid the effects of drought. Aurora City Ordinance 80-47 prohibits new residences from having more than half of their landscaped area in bluegrass; any additional turf has to be planted with a drought-resistant variety with lower water requirements.[1] In other cities, such as Novato, California, homeowners are given a cash incentive for reducing the amount of grass in their yard. In a program called "Cash for Grass," homeowners receive rebates from the local water company if they replace their traditional turf lawn with drought-resistant plants. This puts cash in the homeowner's pocket and reduces the demand on an over-tapped water company.[2] Public agencies can help you choose the right type of grass for your region. For example, the Texas Water Development Board offers a guide which suggests buffalo grass for the driest parts of the state.[3]

Choosing grasses that are inappropriate to the specific conditions of your lawn's environment can lead to an unnecessary environmental impact. If a shade-intolerant species like Kentucky bluegrass is planted beneath a large tree, it will grow poorly. Its poor growth may cause the homeowner to add fertilizer and water, when a better solution would be to put the fertilizer and the water away and plant a grass that can tolerate shade.

Sometimes adding a nongrass species can be advantageous. Clover can enhance a lawn in several ways. Clover has a microorganism that lives in close association with its roots. This microorganism takes nitrogen from the air and converts it to a form that the clover can use. When the clover dies or loses plant parts, nitrogen is added to the soil, where both grass and clover can use it again. This unique arrangement reduces the need for nitrogen fertilizers. Clover also has attractive white flowers, its roots bind the soil, and it can live in harmony with grass.

You might ask why this wonder plant is not a more widely accepted component of today's lawn. Several decades ago, clover was a part of most American lawns. Although some people have never appreciated its bee-attracting flowers, clover officially went out of style when a large seed and chemical company launched a campaign against clover in the lawn.[4] Perhaps it is time to bring clover back.

In the Freedom Lawn, other broad-leaved plants might find a new home. Rose Marie Nichols of Nichols Garden Nursery in Albany, Oregon, and Tom Cooke of Oregon State University have developed ecology lawn mixes.[5] These ecolawn seed mixes, often marketed as "eco-turf," include various grasses, low-growing broad-leaved plants, and flowering perennials that require less maintenance, water, and fertilizer than traditional industrial lawns. Common plants include strawberry clover, dandelion, creeping yarrow and thyme, and English daisy. These growing mixtures result in a three- to six-inch meadowlike lawn that requires a monthly mowing and watering regime.

The use of fossil energy is absolutely essential to the functioning of modern society, but we now know that the by-products of combustion are also at the core of most environmental problems. By using fossil energy more carefully, conservatively, and efficiently, we can begin to reduce the environmental impact of our energy consumption.

We can directly and indirectly decrease the use of fossil energy associated with lawn care through many avenues. First, many gasoline-powered machines are associated with lawn care: mowers, blowers, edgers, mulching machines, and so forth. We can reduce the harmful effect of these machines by using them less frequently or by not purchasing them in the first place. Electric devices are little better, because the pollution is produced at the power station rather than in your backyard. As we saw in Chapter 4, the tiny two-cycle engine that powers most rotary mowers is extremely polluting. Many lawns currently being cut by power mowers could be cut with a nonpowered, hand-pushed reel mower. For a lawn of modest size, a hand-pushed reel mower requires only slightly more effort than a hand-pushed power mower. The relative quiet and the fresh smell of cut grass uncontaminated by exhaust are extra rewards.

A reel mower's cutting action cuts the grass blade straight across rather than at an angle, resulting in less cut area on the grass blade and thus in a lower rate of water lost from the cut. Hand-pushed reel mowers are not effective in very tall grass, however. Manufacturers might take note and develop hand-pushed reel mowers that can handle taller grasses. Longer blades act to shade the soil and to reduce evaporation and root stress. Longer grass usually means deeper, more efficient roots that can better withstand drought and disease.

If you wish to let your lawn grow to a greater height, the gasoline-powered rotary mower may be the best current option. Because rotary mowers tend to rip the blades, which increases water loss, it is a good idea to keep your rotary blade sharp. Frequent sharpening and balancing of your mower blade can reduce fuel consumption by as much as 22 percent. If you use a power mower, adopting a management plan that reduces the frequency of mowing will reduce your contribution to global warming and air pollution. With a nonpowered, hand-pushed mower, the frequency of cuts is not a factor contributing to pollution.

In Chapter 4 we discussed some of the indirect uses of fossil energy associated with lawn care. Substantial amounts of fossil energy are associated with the production and use of fertilizers and pesticides. Minimizing their use conserves energy and reduces pollution. The same is true of irrigation. Similarly, considerable energy is required to move water around, and as we previously discussed, in dry climates the amount of energy used in irrigation can equal that used in mowing. In expressing your concern about the management of your piece of the biosphere and your contribution to global warming and pollution, there is no decision of greater importance than your decision about your lawn's fossil energy needs.

FERTILIZERS

Lawn fertilizers can contribute to water pollution through runoff and seepage into groundwater as well as to air pollution generated by burning fossil fuels. You might consider reducing fertilizer use or eliminating it altogether, as might be done with a Freedom Lawn.

The most familiar fertilizers are the granular fertilizers stocked on garden store shelves with the familiar nitrogen-phosphorus-potassium formulation. Usually these chemicals are relatively soluble and thus can be quickly dissolved by rainwater and carried into the soil.

Organic fertilizers derived from decomposing organic matter can provide for most of your lawn's needs. The nutrients in these fertilizers are largely insoluble and are released over time by microbial action. Some believe that lawns treated with organic fertilizers tend to "bounce back" more quickly after droughts than do lawns fertilized with synthetic chemicals. Organic fertilizers tend to promote deeper root systems. In turn, these longer roots can reach more water, even during times of drought, and thus reduce water stress on the grass plants.[6]

You can create your own fertilizer by leaving your clippings on the lawn. They will act just like the organic fertilizers discussed above by breaking down slowly and releasing their contained nutrients. A thousand-square-foot lawn can produce five hundred pounds of clippings, which, when left on your lawn, can reduce fertilizer use by 20 to 30 percent.

Another source of nutrients for your lawn or garden is compost. All the leaves you rake in the fall or clippings from your shrubbery or any vegetable debris, including vegetable kitchen waste, can be placed in a compost pile in the backyard. With time and an occasional turning, rich black humus is produced. This can be applied to the lawn, shrubs, flower beds, or vegetable garden. It is among the best fertilizers available. By leaving clippings on the lawn and composting, not only are you recycling and saving on commercial fertilizer use, but you are simultaneously helping to reduce your town's solid waste problem.

PESTICIDES AND HERBICIDES

Pesticides constitute a large array of chemicals designed to kill organisms that damage cultivated plants. Many also have the potential to contaminate food chains,

Monitoring

–How: place coffee cans with both ends open, a few inches in the ground, fill with water, if the chinch bugs are in your soil, they will float to the top in about 5–10 minutes.
–When: once a month starting in early spring
–Where: in 4–5 random locations in your lawn

Biological controls

–Once the chinch bugs are at a minimum population level, use insect predators, available to homeowners, to maintain control.

Physical control

–Soap solution and white flannel cloth: spray infected areas with a soapy solution and cover with a flannel cloth; the bugs will crawl out of the ground and stick to the cloth.
–Dispose of the bugs so they do not reinfect the area

Cultural control

–Use slow-release nitrogen fertilizer to maintain correct levels
–Aerate soil (soil compaction results in parched lower layers)
–Plant grass varieties that are resistant to chinch bug attack

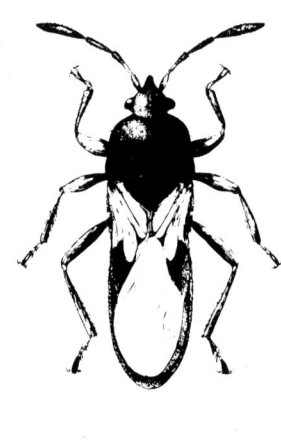

FIGURE 37. The treatment of chinch bugs as an example of Integrated Pest Management. This approach uses a variety of controls applied at critical times in the pest's life cycle to keep its population to a nondamaging level. Control is achieved with a substantial reduction in pesticide use. Photo: courtesy U.S. Department of Agriculture. For further information, see S. Daar, "Safe Ways to Control Chinch Bugs on Lawns," *Common Sense Pest Quarterly* 2, no. 1 (1986): 20–22.

kill nontargeted organisms, and cause human health problems. In general it is best to avoid pesticide use. If pesticides must be used, they should be applied with extreme caution and careful forethought. Read all labels carefully and seek advice from your county agricultural agent. The local hardware store clerk may not be the best source of information. Before you seek advice and service from a lawn care company, find out if they have a reputation for environmental soundness.

Herbicides are pesticides designed to kill weeds, but they can also adversely affect other organisms, such as beneficial soil fungi. Herbicides are often mixed with fertilizers. Weeds are plants that someone or some corporation has decided are undesirable. As we have seen, some plants considered weeds in the Industrial Lawn, such as clover, are a welcome, integral component of the Freedom Lawn. Thus your need for herbicides is strongly influenced by what you conceive to be a weed and the degree to which you wish to eradicate weeds. If you favor the Freedom Lawn or the clover-grass lawn and are willing to tolerate a few "weeds," you will probably not need herbicides.

Be sure that any pest problem really needs attention; many do not. The appearance of a few insects, for example, does not require a cascade of preventive spray. If your problem does require attention, several categories of treatment are available: cultural, biological, and, as a last resort, chemical (figure 37). Cultural controls involve changing procedures or plant cover. For example, overwatering can lead to plant diseases caused by the growth of fungi. Altering your watering regime is easily accomplished and safer than spraying with fungicides. Plants ill-adapted to a site are sometimes attacked by pests. In shady locations, replacing sun-requiring grass with a shade-tolerant species might solve the problem. Under heavily shading trees, replacing all ground cover with thick mulch might avoid disease problems.

Biological controls take advantage of natural enemies, predators, or diseases that afflict a particular pest. One of the best known biological controls is milky spore disease, a bacterium that attacks a wide variety of insect pests.[7] Other examples include predaceous nematodes, which can be introduced into the soil to destroy grubs, and ladybugs, which love to eat aphids.

The combined use of cultural, biological, and chemical strategies is called Integrated Pest Management. Cultural controls are used as a first line of defense against pests and diseases that may be threatening the health of your lawn. These methods are safe and generally easy. Biological controls offer a second line of defense; combined with cultural practices, these controls can solve most of your problems. As a last resort, chemicals may be applied to bring a pest population under control. Once the problem is in check, cultural and biological methods can take over, eliminating the need for continued chemical use.

WATER

Like fossil energy, water is essential to modern society. Without an abundance of water, our society would come to a screeching halt. The water we use comes from precipitation. Most water that we use is fairly recent precipitation that flows in streams and rivers or is stored in reservoirs or rapidly recharged groundwater aquifers. Some water comes from deep aquifers and is a mixture of fairly recent water and older water that entered the ground centuries ago, like the water in the Ogallala Aquifer in the central United States. If the current removal rate exceeds the current recharge rate, the aquifer is on its way to exhaustion.

We can think in terms of how much runoff from the land is needed to meet the per capita use of water. Runoff is the amount of water left after water used by plants and by evaporation is subtracted from precipitation, and per capita use includes not only water used for personal uses but also the water used by industry, electrical energy production, commercial cooling, agriculture, lawn irrigation, fire fighting, and street cleaning. To meet the per capita water needs of a person living in an arid region with relatively low rainfall and relatively little runoff, a vast land area is needed. For example, the water needs of Los Angeles are drawn from as far away as northern California and Colorado. New York City, on the other hand, because of its greater rainfall, draws its water supply from about the eastern third of the state. In spite of the abundance of rainfall in New York, New York City, like Los Angeles, is deeply concerned about its water supplies. At various times, the reservoirs that supply New York City and that usually refill during the winter months are far below expected levels, raising the potential of serious shortages during the hot summers.

The dams and reservoirs built to meet the water needs of our extensive urban areas represent some of the great engineering accomplishments of our age, but the potential for increasing urban water supplies without further damaging aquatic environments is not great. Because we cannot easily or cheaply increase our environmentally sound water supplies, the best course of action is to conserve water by using it more efficiently and by minimizing nonessential uses. Throughout the country we see many evidences of this policy of water conservation.

Lawn irrigation can be considered a nonessential use, and in humid regions many lawns flourish without any irrigation. As mentioned earlier, saving water may be a matter of selecting drought-resistant grass varieties or tolerating some brown in your lawn, knowing that once rainfall returns the lawn will green up. Sometimes irrigation may be needed; when a new lawn is being established, for example, the soil should be kept moist. In making the transition away from an Industrial Lawn, some may wish to continue irrigating but may choose to do so more efficiently and more modestly. This requires following a few simple rules.

- Apply only as much water as needed. Studies have found that many homeowners apply twice as much water as lawns need.[8] Residents participating in a study in Logan, Utah, watered their lawns with little regard to the requirements of the grass.[9]
- Know how much water your sprinkling system delivers per hour.
- Have a good idea of how fast water will sink into the soil and do not apply water at a faster rate.
- Water in the evening.
- Water less often but deeply rather than more often but shallowly.

Root growth depths with different watering regimes are shown in figure 38. Much of this can be learned by personal experimentation, but you may wish to seek the advice of a lawn care professional known to be interested in water conservation.

Another strategy for reducing water use in your yard calls for clustering plants

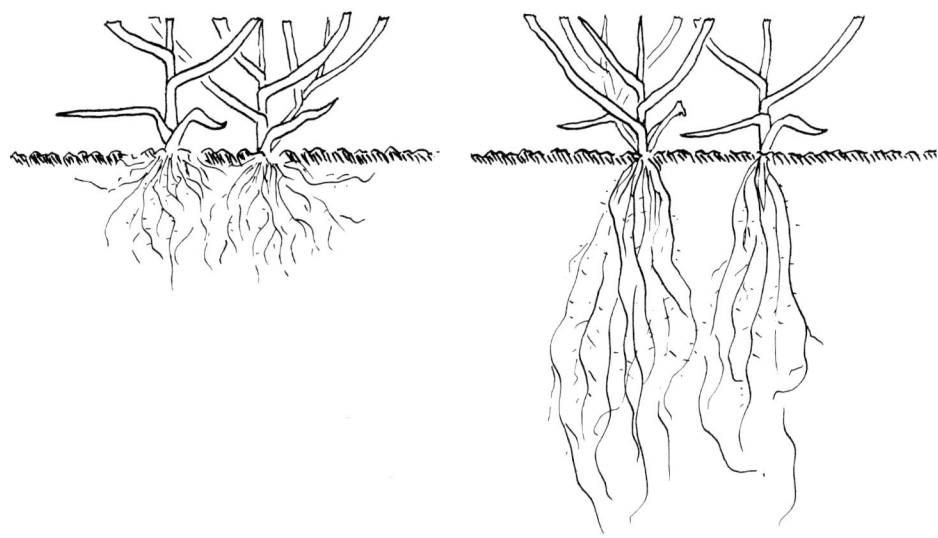

light watering   thorough watering

FIGURE 38. Your watering regime can affect a grass plant's root system and its response to drought. Frequent light watering will produce shallow roots, making the plant highly susceptible to drought. Less frequent but deep watering produces deep roots that are able to withstand modest drought. Drawing © Lauren Brown.

with the same water demands in the same place or hydrozone. This method, called hydrozoning, allows you to more easily match water applications with the water requirements of plants.[10]

Yet another conservation method involves using so-called gray water from showers, washing machines, and sinks in your yard as long as it is not prohibited by law. In one home in Pebble Beach, California, a gravity system uses a five-hundred-gallon tank to collect water from the roof and shower for use in the garden.

However you do it, reducing the amount of water used to irrigate lawns can make a noticeable contribution to the conservation of water supplies and, indirectly, to a reduction in our use of fossil energy.

CHOOSING A LAWN CARE COMPANY

Although many people enjoy yard work and look forward to an afternoon spent outside, 22 million Americans would just as soon leave these tasks to someone else.[11]

Today, the service provided by lawn care companies appears to be affordable and simple. Many companies have reputable lawn care practices and can provide a homeowner with a carefree, healthy lawn. Other companies, however, can lead you down the road to what might be called the lawn care equivalent of drug addiction, where to achieve the continuously green lawn you must repeatedly apply chemicals, which requires that you must add water, which then requires more chemicals, more water, more chemicals, more water, ad infinitum.[12]

When hiring a lawn care company, remember that your lawn is your property, your little piece of the biosphere, and that you have the final word. Two things should be kept in mind: greenscam and overselling. Greenscam is a tactic used by unscrupulous companies who, through false or incomplete labeling, attempt to use your concern for healthy, environmentally sound products to get you to buy products that have no particular health or environmental virtue and may, in fact, have negative side effects. Overselling is the natural tendency of any salesclerk who wishes to increase sales.

Some lawn care companies have the right environmental rhetoric but follow practices that do not accord with environmental objectives. You are told one thing, but the actuality is quite different. Other companies may try to persuade you that a pesticide application is necessary when you have no pests, or they may advocate two or more fertilizer applications where none or one would do.

In choosing a lawn care company, whether it be a nationally recognized chain or a local concern, you must distinguish between businesses that will make environmentally sound decisions about your lawn and those with other objectives such as selling you an Industrial Lawn or selling some environmentally unsound practices under the guise that they are sound. The best way to make intelligent decisions about lawn care companies is to know your lawn care objectives and the management alternatives you would adopt if you were caring for your lawn yourself.

THE FREEDOM LAWN

One of the best ways of changing your management practices and reducing your environmental impact on the biosphere is to switch to a Freedom Lawn. In humid regions, if you abandon or reduce the strict demands of the Industrial Lawn for frequent fertilization, applications of various pesticides, frequent watering, and rigorous mowing, your lawn will gradually become a Freedom Lawn. Bare spots may develop, but since the lawn is under a continuous bombardment of seeds from nearby herbs, shrubs, and trees, new plants will take hold, their presence determined largely by their ability to survive below the height of the mowing blade. Not only is it likely that the number of species will increase, but also the numbers of indi-

viduals in some species will increase. This applies not only to green plants but also
to animals in the soil and microorganisms whose presence will respond to the cessation of chemical applications. Changes in soil tilth and aeration might be expected
as soil organisms become more active. Better drainage might result. The lawn might
shift back toward the more natural state, powered primarily by solar energy, watered
by rainfall, and supplied with nutrients found in the soil. It would be more disease
resistant because of its diversity of species. Growth rates would fall with the cessation
of fertilization and irrigation; less frequent mowing would be required. The environmental impact of your lawn on air and water quality would be sharply reduced, and
biodiversity would increase.

Although the Freedom Lawn is already the lawn of choice for many homeowners, its acceptance by homeowners for whom the Industrial Lawn is still the ideal
will require a major change in attitude, a serious attempt to see things differently.
Dandelions and crabgrass might become things of beauty and admiration, and brown
spots would be evidence of natural cycles.

Reducing the Amount of Lawn in Your Yard

You might ask yourself whether you need all of your lawn. Are there areas of
your yard where grass is not needed? What about steep slopes, heavily shaded areas,

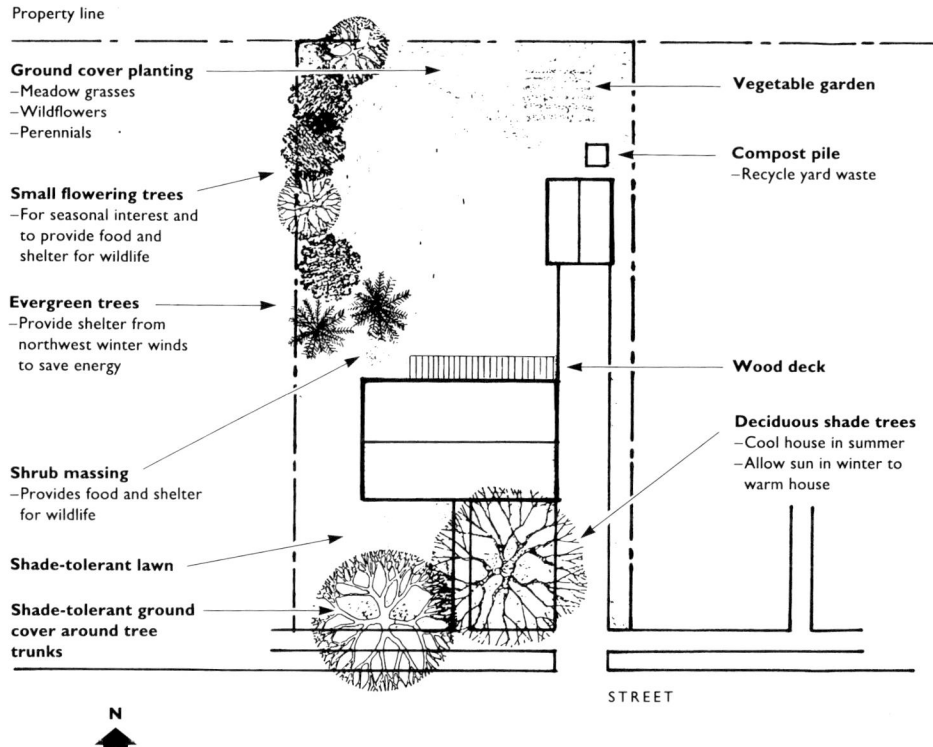

REDUCING THE LAWN'S PROPORTION

Property line

Ground cover planting
–Meadow grasses
–Wildflowers
–Perennials

Small flowering trees
–For seasonal interest and
to provide food and
shelter for wildlife

Evergreen trees
–Provide shelter from
northwest winter winds
to save energy

Shrub massing
–Provides food and shelter
for wildlife

Shade-tolerant lawn

**Shade-tolerant ground
cover around tree
trunks**

Vegetable garden

Compost pile
–Recycle yard waste

Wood deck

Deciduous shade trees
–Cool house in summer
–Allow sun in winter to
warm house

N

STREET

FIGURE 39. Reducing the lawn's proportion of the yard: replace parts of the lawn with small trees, shrubs, flowers, or ornamental grasses. A modest change of this kind can create a more varied landscape as well as reduce the lawn's impact on the environment.

corners of your property, heavily traveled paths? Replacing the lawn with low-maintenance herbs, shrubs, or trees reduces the need for any water or chemicals you might be adding to the lawn and for the need to mow (figure 39). In some parts of the yard, gravel, wood, or brick patios or decks might make better sense than a lawn. Of course if you are building a new home, you may choose to preserve some of the original vegetation and blend it with lawn.

Replacement alternatives from the plant world are nearly endless. You should, however, choose plants adapted to where you live that can maintain themselves without added supplements and are appropriate to the size of your yard. These may be plants from your native region or introduced plants that have naturalized to local conditions. Such plants are not hard to obtain; native plant nurseries and mail order houses are scattered throughout the country. Many are listed in *Gardening by Mail,* by Barbara Barton, and in Henry Art's six-book series, *The Wildflower Gardener's Guide.*[13]

Trees and shrubs can be used to introduce both structural variety and species diversity and to attract birds and other wildlife by providing food and shelter. Plants might be used as architectural elements to create spaces or provide privacy. Plants can also be used for glare reduction, traffic control, and sound abatement. Shrubs can act as a snow fence. Well-placed trees can modify the flow of solar energy and wind and reduce a household's cost for heating and cooling.

Many homeowners might consider using part of their lawn for vegetable or flower gardens. There can be great satisfaction and pleasure in growing your own vegetables. They taste better, and you can be certain that they do not contain any chemical residues. Flowers add beauty to the yard and can be cut for bouquets to bring the beauty of the yard into the home.

Brick terraces, sculptures, fences, and potting sheds are often used as elements of the ground plan. These elements emphasize line and shape. In addition to their aesthetic value, many surfaces are both practical and colorful: think of bluestone, bricks, granite, gravel, and wood. They can be beneficial to the natural ecology of the yard: when properly spaced, these materials allow water to soak into the underlying soil.

All across the country people are experimenting with changing the composition of their yards and introducing native species. Start with a small area and gradually reshape the whole property. In that way, your neighbors will have time to learn what you are trying to do and may even follow suit.

Replacing the Lawn

Consider taking the idea of reducing the amount of your yard planted in grass a step further: consider replacing the entire lawn. The current popularity of the lawn is closely tied in to its beauty, first forged by a handful of artists in eighteenth-century England; new concepts of yard use should not only respond to our environmental and ecological concerns but present equally powerful aesthetics. Can these alternatives be beautiful, as beautiful as our lawn? In response to this question, we present a series of images (see figs. 40–46) from different regions and cultures that can serve as a point of departure for thinking about the beauty of landscapes that are *not* based on the lawn.

If you wish to replace your lawn, it is important to decide what you want to achieve with your yard ecologically, aesthetically, and functionally. If you do not have the time or inclination to do the work yourself, you might choose a landscape architect, horticulturist, or artist who can give form to your new goals. Before hiring

FIGURE 40. A Mediterranean courtyard. This southern European residential landscape has been adapted with great success to southern California, which also has a Mediterranean climate, and with moderate success to the desert Southwest, which is drier. Photo: © Jonas Lehrman. Reproduced with permission from Jonas Lehrman, *Earthly Paradise: Garden and Courtyard in Islam* (Berkeley: University of California Press, 1980).

FIGURE 41. Although Japanese gardens are aesthetic responses to historic, cultural, and geographic conditions that differ from our own, they illustrate a well-designed landscape devoid of grass and accomplished with great economy of means. We can take lessons from their minimal use of resources, low consumption of water, surface drainage, and low maintenance in terms of chemicals, insecticides, and mechanical upkeep. Ko-myo-ji, Japan. Photo: Diana Balmori.

one of them you need to know if they are responsive to your ideas and to the issues raised by this book. Ask for examples of their work and for a description of how they would implement your goals.

USING IDEAS FROM OTHER CULTURES

Landscapes of great power and beauty not based on the lawn have evolved in other times and climates. Although many cultures surround the Mediterranean Sea, the basin shares a relatively dry climate. One remarkably consistent landscape has emerged there: paved courtyards surrounded by walls and lavishly planted with vines and potted plants, where pitched roofs direct water from infrequent rains into a small pool in the courtyard center. The walls surrounding the yard provide shade and protect against drying winds. The vines planted along the walls as well as the multitude of pots fill the patio with the scent of flowers and herbs and shade the walls and ground from the sun's heat (figure 40).

The Japanese landscape tradition is derived from the Chinese, but the influence of Zen Buddhism took the Japanese temple gardens in a much more abstract direction. Gardens, devoid of any grass or planted beds, contain crushed stone or raked sand (which allow infiltration of rainwater), rocks, shrubs, trees, and a path that must be carefully followed. Groupings of shrubs and trees are skillfully layered, and beneath large trees moss can sometimes be found. The temple tradition in gardens was adapted to Japanese house gardens, which are mainly long, narrow corridors, and the careful lessons in scaling down the elements of the garden create an extraordinary sensation of ample space (figure 41).

In the medieval Western tradition, gardens were located behind castle and cloister walls, where protected space was at a premium. There, baskets of woven vines or stone retaining walls enclosed highly productive raised planting beds. These raised beds could be reached from both sides for planting, weeding, or cultivating and were dedicated to specific plantings: medicinal herbs, vegetables, and flowers.

Is there anything in these examples for our time? Perhaps nothing, if we take them literally; in creating new aesthetics, ones suitable to our climatic conditions and our environmental goals, however, we can extract from each universals that offer wise instruction.

Today, in some of our warmer climates, much can be learned from the climatic control Mediterranean courtyards obtain with walls and trellises and plantings. The abstract crushed stone gardens of the Japanese landscape have a kinship with the environmentally sound stone and cactus yards of the Southwest. The raised planting beds of medieval origin, so easily managed and of great productivity, offer environmental advantages for small spaces or hostile environments because of their efficient use of water and nutrients; they have become the preferred form for modern urban community gardens.

In the 1920s and 1930s, across the United States, several attempts were launched at making local material the source of regional design and identity. The Arts and Crafts Movement of the late nineteenth century, which developed in England in opposition to the homogenization brought about by industrialization, stimulated interest in the regional and vernacular characteristics of place. In the Midwest this art movement took the name Prairie School, and it produced a rich heritage of both architecture and landscape design. In California and the Southwest, it was called the Mission Style and was based on the architecture and landscape of the old Spanish missions and their courtyard gardens. Important local variants existed, especially in New Mexico. There local adobe-building traditions, both Indian and Spanish,

FIGURE 42. In the Southwest, adobe buildings and earthen courtyards with local desert plants were responses to the ideas of the Arts and Crafts Movement of the early 1900s. Later, water scarcity and legislation restricting water use in several municipalities reinforced the earlier movement, producing a landscape with an aesthetic that differs markedly from that of the lawn. Homeowners here landscape their yards with drought-resistant desert plants, crushed stone, and sands of different colors. These landscape designs make little or no demands on local water supplies. Photo: Diana Balmori.

were brought back into favor, as was planting in accordance with the desert conditions of the region. With the increasing stress on water supplies, xeriscapes or desert plantings received another boost.

A PARTNERSHIP WITH THE REGION

A sense of place arises from our ability to recognize where we are in this world by the natural landforms and native species. Many areas of our country have lost their unique identity through the introduction of the Industrial Lawn and nonnative

FIGURE 43. Many city dwellers have uprooted most or all of their lawns to plant shrubs, flowers, and trees. Photo: Diana Balmori.

ornamental species. The reintroduction of native plants can help re-create a sense of place reminiscent of our predeveloped landscape. Removing the lawn and some of the nonnative plants in the yard and replacing them with native plants and native vegetation patterns is one way of reestablishing a local sense of place.

Like Capability Brown's vision, which was so well suited to the English environment, but, when applied elsewhere, was so often environmentally unsound, there is no single environmentally sound design that can be applied everywhere in the United States—our country is just too large and too climatically and vegetationally diverse for one solution to work everywhere. Each region has an array of native species and natural vegetation that is able to grow, survive, and reproduce using the solar energy, rainfall, and soils that the place has to offer. In our search for landscape paradigms that will both meet our environmental needs and satisfy our search for new aesthetics, the best advice may be to look around us.

FIGURE 44. Colonial gardens in New England were often formally laid out in planting beds of herbs, vegetables, and flowers surrounded by fieldstone or plank walks. The entire garden was enclosed by a wooden fence. Many modern restorations of this traditional form offer an alternative to the lawn. Thomas Hyatt House, Ridgefield, Connecticut, designed by Patricia O'Donnell, ASLA. Photo: © Diane Nowotarski Hobé.

Rather than embark on a lengthy discussion of these new locally based aesthetics, we have chosen a few images (figures 44 through 46) to depict aesthetic adaptations from various regions of the country that use nongrass landscapes.

We hope this book will inspire many Americans to reexamine their use of that piece of the biosphere entrusted to their care. By choosing aesthetically pleasing but environmentally sound alternatives to the classic American lawn, we can unite our environmental concerns with direct personal action. Indeed, we can draw the line on environmental degradation in our own yard. Understanding where the lawn's popularity comes from, how the lawn fits into the global environment, and, finally, what changes we can make to alter its effects gives each of us the power to improve

FIGURE 45. In areas with abundant rainfall, such as the Pacific Northwest, under native trees moss can be an alternative to grass. This photo of an area of Irish moss was taken at the Bloedel Reserve, Bainbridge Island, Washington, and was designed by Richard Haag, ASLA. © 1992 Mary Randlett.

FIGURE 46a. In the Southeast, homeowners may choose to keep their yards wooded in native species, thereby eliminating the need for mowing, applying fertilizers and pesticides, and irrigating. Photo: Frank Golley.

FIGURE 46b. The home of the Abu-Ghazalehs in Lead, South Dakota, showing native grasses one year after planting. This vegetation is expected to mature into prairie, dependent on natural rainfall and temperature conditions and two annual mowings. Photo: © Nikki Lee, Earthworks.

our piece of nature. We need not cease to love the lawn. By understanding how it does and does not work, we can adapt it to our time.

6 The Lawn and Sustainability

Possibly the greatest concern facing humanity at the beginning of the twenty-first century is the question, Is the society we have fashioned sustainable, can it be depended upon to fulfill the material and ethical needs of future generations, or does it bear within it the seeds of its ultimate destruction? This question is being vigorously debated in all sectors of society and in all countries throughout the world, and it is the core topic of this book. Our search focuses on finding new landscape designs in closer harmony with nature that also reflect concern for the future; designs based on understanding the functions of nature and the importance of preserving those functions to ensure the future health of human societies.

Many activists are trying to solve environmental, economic, and social problems in the United States. Among them are numerous individuals and groups concerned with Industrial Lawns and the search for new landscape designs—aesthetically pleasing, economically solid, environmentally sound—that contribute to long-term sustainability.

In this closing chapter, we present examples of communities, universities, corporations, developers, and governmental agencies throughout the nation who are exploring and implementing environmentally friendly lawn designs and management techniques to replace those of Industrial Lawns now common on millions of acres of public and private lands. These new landscapers start with the recognition that lawns are part of a larger ecosystem, the earth, and in the process of redesigning the American lawn, they are creating new aesthetics, employing different economic scenarios, elevating the role of natural processes in the design process, and formulating new educational goals where practice matches theory.

ADOPTING THE FREEDOM LAWN

Earlier we described homeowners who, dissatisfied with Industrial Lawns, sought alternative landscape designs, usually some form of the Freedom Lawn. Their motivations varied from wishing to spend less time mowing and an enthusiasm for experimentation to desiring a new aesthetic and increased environmental consciousness. Whatever their reasons, these homeowners acted independently and without community support. Imagine what would happen if their actions were multiplied in a neighborly fashion and encouraged at the community level. What would be the result?

A Freedom Lawn Initiative

In Milford, Connecticut, a town that overlooks Long Island Sound—one of twenty-eight estuaries designated of national importance—many homeowners have good reason to be proud of their yards and their community. Each year, additional Freedom Lawns dot the suburban landscape and more residents become aware of alternative lawn care and landscaping techniques. The credit for this change goes to the town's ad hoc environmental committee, the Environmental Concerns Coalition (ECC), based in the mayor's office.

In 1990, several Milford citizens formed the environmental coalition as a response to Earth Day. The advocacy group wanted to increase the ecological awareness of citizens and to help them become more environmentally responsible neighbors. In 1996, ECC adopted the Freedom Lawn concept as a way to encourage voluntary antipollution measures. This focus made sense because Industrial Lawns were an important source of chemicals that contributed to the pollution of Long Island Sound.

Hundreds of leaflets placed in town stores and on car windshields announced the coalition's call to action. The text briefly described the sources of lawn pollution and how people could change their lawn care management regimes. After a flurry of phone calls from interested citizens, a more detailed brochure proposed the formation of the Milford Community Freedom Lawn Initiative.[1] This initiative suggested that an effective way to raise environmental awareness would be through an annual Freedom Lawn competition that would give communitywide recognition to Freedom Lawn owners for being neighborhood role models and educators.

BE EVER MINDFUL OF OUR FRAGILE ENVIRONMENT • SAVE THIS PRECIOUS PLANET FOR OUR CHILDREN

FREEDOM LAWN

"COMMUNITY FREEDOM LAWN INITIATIVE"
PRESERVATION BEGINS IN YOUR OWN BACK YARD
A JOINT PROJECT OF
ECC / ENVIRONMENTAL CONCERNS COALITION
& THE MILFORD CONSERVATION COMMISSION

Reproduced with permission from the Milford Freedom Lawn Initiative. Designed by Robert Proto.

The competition is based on homeowners' landscape designs and lawn care management techniques. Owners or their neighbors nominate yards as Freedom Lawns, which ECC volunteers then evaluate on the basis of minimizing local and regional environmental damage and increasing biodiversity. Judges look for chemical-free lawns that contain a diversity of plant species, patches of forest or meadow, bird feeders and birdbaths, and the presence of gravel driveways. (Gravel driveways allow rainwater to sink into the soil rather than run off into the street. This has several benefits, among them providing water for plant growth and reducing storm flow.) Each year, up to five winners selected from more than fifty nominees receive a ceramic butterfly, locally handcrafted and donated to the program, a decorative sign for the lawn, and recognition in the annual Freedom Lawn Initiative booklet and in local newspapers.

This creative approach of encouraging homeowners to connect their landscaping decisions to consequences beyond their property boundaries is growing in popularity. In 1998, two thousand copies of the Freedom Lawn booklet were published. A variety of local organizations assisted with the printing costs, including the mayor's office, the Inland Wetlands Agency, the Milford Conservation Commission, a high school ecology club, and several private businesses. Two years after the first public announcement, it was clear that the community had accepted the Freedom Lawn Initiative.

Encouraged by the success each year brings, the ECC continues to build new aspects into the competition. In 1997, the ECC organized a tour of the winners' yards. Tour participants received brochures and advice from homeowners concerning low-impact lawn care regimes. The following year, a Freedom Lawn Directory was created that enabled the public to visit and evaluate Milford's one hundred Freedom Lawns. In 1999, Freedom Lawn streets and neighborhoods, defined as areas that contain 50 percent or more Freedom Lawns, were identified. Through speaking engagements, ECC volunteers and Freedom Lawn winners have spread knowledge of the initiative throughout the state, and the Milford brochure is frequently cited in newsletters of other towns. Now the ECC plans to work with garden clubs and other groups that promote chemical-free yards throughout Connecticut and to have a statewide Freedom Lawn Day.

If the citizens' activity in Milford were replicated in towns across the nation, our 21 million acres of lawn would look quite different. There would be less air and water pollution, a greater variety of plants and animals, more visual stimulation, and less fear of lawn care chemicals.

THE HOMEOWNERS' VIEW

Many Milford Freedom Lawn owners believe that their quality of life has improved. For them, maintaining a Freedom Lawn has become an extension of an overall way of living. One Freedom Lawn winner believes that "saving money and purifying the ecosystem most of all is a good feeling, even if we're cleaning it up only in a very small way." Says another, "Before we switched to organic methods, we had gardens by Ortho with less and less production from more and more chemicals. The day the boys and I couldn't find worms in the garden for fishing in the river was the day we decided to try what organic gardening was selling. The food and flowers we produce may save us a little compared to what we would spend in a supermarket, but the quality of the food is vastly superior. And what would we have to pay for the beauty?"

James Sterling believes that his family gains multiple benefits from maintaining a Freedom Lawn (figure 47). The money he saves by maintaining a more natural yard and not using pesticides goes into his children's college education fund. He asks, "Why should we put ourselves at risk with a chemical lawn? I love knowing that my kids can play in the garden, eat the herbs and be safe."

The annual Freedom Lawn brochure provides a brief description of each winner's property. Here are some excerpts from various years:

• "The look of a wild flower garden, a clever asymmetry of the actual proportions of the planted front yard beds . . . a shorebird and perhaps

FIGURE 47. April and James Sterling, 1998 Milford Freedom Lawn Initiative winners, with their three daughters outside on their Freedom Lawn enjoying the many things of interest, such as earthworms, insects, and flowers. Reproduced with permission from the *Connecticut Post* © 1999; photo: B. K. Angeletti.

other midsummer creatures hiding among the blossoms make us smile at the little brown shingle house. . . . Somewhere a lawn mower must repose, rusting and abandoned" (1999)

• "This Freedom Lawn will never be sued for lack of affordable housing— for birds! Two lovely smaller maples, each containing at least 10 feeders, cast shade onto the front yard. Wide, brick-edged beds of flowers on each side . . . beds at the base of the bird bath" (1999)

• "A poster for the neighbor-friendly Freedom Lawn idea: nourish the birds, nourish the eyes, enrich the neighborhood with plants and flowers selected to sustain color vibrancy from season's start to finish" (1998)

• "Crowned by the queen of the butterfly kingdom, catalyzing the area by its bold example, this fountain of flowers seems posed to take over the side street as its next beauty conquest. Three seasons are distinctly marked. Slopes and banks cascade down in color ripples to levees of flower boxes which barely contain the flood" (1998)[2]

A number of organizations encourage homeowners to rethink their attachment to the Industrial Lawn. We hope that by mentioning a few of them, we can help strengthen this environmental message and heighten public awareness.

The Garden Club of America has the potential to create one of the largest environmental impacts on the American public. The seventeen-thousand-member organization has 194 clubs in forty states and the District of Columbia. One way the Garden Club helps improve environmental quality is through environmental education. Its pamphlet *The New American Lawn* advocates environmental responsibility among homeowners and corporations, and essentially proposes a Freedom Lawn.[3]

As part of its public outreach program, the Rhode Island Wild Plant Society has distributed buttons with the motto "Why Mow?" stamped above a lawn mower with a red line crossed through it. A biannual newsletter contains articles that include alternatives to lawns and tips on drought-hardy plants (figure 48).

Some organizations are challenging the Industrial Lawn solely from a human-health standpoint. A cause-and-effect relationship has been found between different cancers and pesticides. For this reason, the American Cancer Society publishes a brochure called *Drug Free Lawn*. Its purpose is to educate the public about alternative, less-toxic ways to control insects and weeds in the yard.

The Grassroots Coalition lobbies for legislation to reduce pesticide use in and around schools and educates the public about the hazards of such chemicals. This Connecticut-based organization networks with other environmental groups and publishes a descriptive brochure called *Neighbor to Neighbor: Useful Information for a Beautiful Landscape*. Underlying reasons for reducing pesticide use are provided, along with tips on how to troubleshoot one's yard, to use Integrated Pest Management, and to have a chemical-free lawn.

REDESIGNING EDUCATION: THE UNIVERSITY CAMPUS

The connection between teaching ecological principles and conservation and the facilities and landscape management of college and university campuses went unrecognized for many years. Owing to a lack of communication among administrators, managers, and faculty, environmental principles of conservation and care taught in the classroom were ignored in daily campus operations.

Inefficient lights burned in empty rooms, windows remained open in overheated rooms, water dripped from leaky faucets and ancient toilets, old electric motors

NEWSLETTER
March 1999, Vol. 13, No. 1
P.O. Box 114, Peace Dale, RI 02883-0114 (401) 783-5895

Why Mow?

"Rethinking the American Lawn"

**The gardens of RI's future will be more environmentally friendly,
as gardeners use more native plants, less water,
and fewer toxic pesticides and fertilizers.**

Be sure to come see the RIWPS Display Garden & Booth
and other "Gardens of the Future"
at the
Sixth Annual Rhode Island Spring Flower & Garden Show
February 18 - 21, 1999
at
Providence's RI Convention Center

FIGURE 48. The Rhode Island Wild Plant Society (RIWPS) produces a biannual newsletter that helps promote their Freedom Lawn efforts. The society's "Why Mow?" garden display at a 1999 state spring flower and garden show received a prestigious award. Reproduced with permission from the Rhode Island Wild Plant Society. Designed by Gretchen Halpert.

wasted many kilowatts of electricity, and groundskeepers poured bags of fertilizers and pesticides and gallons of water onto landscapes in pursuit of the Industrial Lawn. These neglectful practices made many of the nation's four thousand universities and colleges major contributors to air, water, and noise pollution.

David W. Orr, professor of environmental studies at Oberlin College in Ohio, a leader in the movement to connect the campus and the classroom, believes that institutions need to act with greater environmental responsibility. Students' educational experiences can be enriched when their university addresses its environmental problems, writes Orr. Through their daily operations and in their management of the campus environment, colleges can teach students and the surrounding community to act as responsible citizens, which will benefit not only the environment but also the community and the economy.[4]

Many educational institutions are today exploring ways to link the classroom and the campus. In 1994, an international conference at Yale University brought together 450 faculty, staff, and student delegates from around the world. The result of this campus summit was the publication *Blueprint for a Green Campus,* which demonstrated how higher educational institutions could contribute to an environmentally sustainable future through better campus design and management. Another example is the National Wildlife Federation's Campus Ecology Program. Program personnel work with faculty, students, and managers to promote and implement conservation programs on campuses throughout the country.

Here we focus on university lawns and landscapes, presenting strategies designed to teach stewardship and to protect natural resources and human health. Not only do these strategies connect the classroom to environmentally sound management of educational landscapes, but they also provide models for students who will eventually manage the millions of acres of lawns under the control of institutions and businesses and local, state, and federal government.

University Models

Institutions such as Penn State University, Oberlin College, the University of Arizona, the University of Vermont, Connecticut College, the University of Wisconsin–Madison, and Brown University are seeking to connect word and deed by emphasizing environmentally sound campus aesthetics, energy efficiency, waste reduction, and concerns for human health. These goals meet the needs of the institution and bring students, professors, and staff into the decision-making process and have the expectation of achieving economies through a more judicious use of resources. Landscape reforms at four varied educational institutions show how this new approach can work.

To address the need for cooperation and responsible leadership, in 1990 an international conference on the role of universities in environmental management and sustainable development was hosted in Talloires, France. The resulting declaration has become the guiding document for the Association of University Leaders for a Sustainable Future. More than 260 institutions of higher education in more than forty countries across five continents use this document as a template. The text outlines the appropriate role and responsibilities of universities for supporting sustainable development and taking the lead in environmental literacy. Excerpts from the declaration follow:

> We, the presidents, rectors, and vice chancellors of universities from all regions of the world, are deeply concerned about the unprecedented scale and speed of environmental pollution and degradation, and the depletion of natural resources.
>
> Local, regional, and global air and water pollution; accumulation and distribution of toxic wastes; destruction and depletion of forests, soil, and water; depletion of the ozone layer and emission of "green house" gases threaten the survival of humans and thousands of other living species, the integrity of the earth and its biodiversity, the security of nations, and the heritage of future generations. . . .
>
> We believe that urgent actions are needed to address these fundamental problems and reverse the trends. . . .
>
> Universities have a major role in the education, research, policy formation, and information exchange necessary to make these goals possible. Thus, university leaders must initiate and support mobilization of internal and external resources so that their institutions respond to this urgent challenge.

CONNECTICUT COLLEGE

In 1997, Connecticut College in New London, Connecticut, signed the Talloires Declaration (see box).[5] As part of its pledge to conservation and to a sustainable environment, the college "acknowledges its responsibility to teach environmental stewardship, not just in the classroom, but in all campus operations as well. By striving to make operations more efficient and environmentally-sound, Connecticut College can serve as an environmental model while also saving money and resources."[6]

The goals of the college's program are to preserve open spaces that enhance human health and well-being, to protect habitat and increase biodiversity, to save energy, to extend the life of renewable and nonrenewable natural resources, to curb air, water, and soil pollution, and to minimize waste. The grounds of this small liberal arts college campus are managed as a 750-acre arboretum, and the entire landscape is considered both an integral part of the teaching environment and an outdoor laboratory.

UNIVERSITY OF NEVADA AT LAS VEGAS

This campus, located in the arid West, provides a good illustration of student initiative in saving natural resources. In 1991, undergraduate Tara Pike started a student-run organization called Students Conscious of Protecting the Environment (SCOPE), initiating a new approach to landscaping that took into account aesthetics, ecology, and a limited budget. The students' first endeavor, Project Desert Landscape, sought to reduce the area of turf on campus by converting grass to a more sustainable, less resource intensive habitat—a desert landscape.

Out of 335 acres on the University of Nevada's campus, 80 acres were devoted to landscape and 59 of those were turf. The university's School of Design and Landscape Architecture proposed that 18 acres of difficult to maintain turf be converted. SCOPE raised funds that were matched by local businesses and donated time and labor.

Today, 10 of the 18 acres have been converted to desert landscapes by planting native southwestern plants, such as the Texas ranger, deer grass, creosote bush, and desert willow (figure 49). Although the grounds department acknowledges that the water used for landscape irrigation is not metered separately for each area, the turf conversion is estimated to have reduced water usage by about 30 percent. This reduction is significant in the arid Las Vegas Valley.

THE UNIVERSITY OF ARIZONA

Support for water conservation is high in Tucson, a city that depends solely on groundwater from deep aquifers.[7] Since 1984 the University of Arizona, home to thirty-five thousand students, has been implementing a turf reduction program. The initial impetus for reducing turf acreage was water conservation, but the use of xeriscapes and desert landscapes also made sense in high pedestrian traffic areas and around new buildings. Eight of sixty acres of turf have been eliminated, saving an estimated 6 million gallons of water a year.

Changing the watering regime achieved further reductions. The irrigation system was upgraded from flood irrigation to a computerized sprinkler system, saving about 9 million gallons of water a year. The grounds department plans eventually to convert the entire irrigation system over to reclaimed water instead of depleting the underground source of drinking water.

SEATTLE UNIVERSITY

At Washington's Seattle University, Ciscoe Morris has managed the fifty-three-acre campus in an ecologically friendly way since 1978. Several years ago, when Morris saw students sunbathing next to pesticide notification signs, he knew a dramatic change was needed. His new management regime uses Integrated Pest Management, with a 99 percent reduction in pesticide use. The new maintenance program has created some dissatisfaction, however, because dandelions and clover appear in the ten acres of campus turf. Small amounts of pesticides of low toxicity are occasionally used on the three acres of athletic fields. In nonturf areas, Morris depends mainly on aggressive ground covers to outcompete weeds. An important part of Integrated Pest Management is to use plants that are adapted to site conditions.

To save water and mowing time, Morris employs a low-maintenance grass in naturally dry areas. To reduce the waste of water, a computer regulates the sprinkler system and adjusts the amount of water used by measuring the need against the natural transpiration rate.

FIGURE 49. An example of lawn conversion is seen in front of John Wright Hall at the University of Nevada at Las Vegas. *Above*, a lawn-dominated landscape, 1993; *right*, the same area converted to a desert landscape, 1999. This change reduced the consumption of water, fertilizers, and pesticides. Photos: Tara Pike.

Morris's determination to create a landscape with limited artificial inputs has the enthusiastic support of the administration, students, and faculty. The grounds are cited as one of the "twenty-three best things about Seattle" in the university's public relation materials. The grounds have also received an accolade from the Washington State Department of Wildlife, which declared the campus a wildlife sanctuary, the only campus in the state to have earned this recognition.

CORPORATE RESPONSIBILITY

Educational institutions may be greening their campuses, but environmental leadership is needed at other institutions, such as corporations and governments. In the competitive world of business, not only must products and services compete for the consumer's attention, but the image projected of a corporation's goals and ideology influences a consumer's purchase decisions. For decades, corporations have endorsed environmental goals publicly, but only recently have such endorsements been recognized as an enhancement of profitability at the organization and corporate planning levels.[8] This marks a shift from the old industrial model to a more environmentally sustainable, economically progressive model.

Like the broad range of environmental efforts that are occurring on university campuses, corporate environmental goals vary from energy efficiency and reduc-

ing facility emissions to recycling and reusing product parts and to maintaining environmental landscapes, as in the examples that follow.

CIGNA CORPORATION

In the early 1990s, American corporations were tightening their budgets by decreasing their expenditures and downsizing their workforces. For many years, CIGNA Corporation, in Bloomfield, Connecticut, maintained 300 acres of well-manicured lawn, requiring a significant investment in labor, machines, chemicals, and water.

Two employees identified a way for the company to save money by reducing landscape costs. Grounds supervisor Tim Sanderson and his supervisor Brian McDonald were familiar with a different approach to lawn care maintenance and wanted to try it.[9] They suggested to CIGNA's top management that both the company and the environment could benefit if some lawn was converted to more natural, less labor-intensive, and less chemical-dependent vegetation, such as meadows and wildflower fields. At the time, this seemed like a radical idea, but it was accepted as an ecologically sensitive and economically sensible approach. CIGNA decided that this new management technique would enhance its corporate image.

The initial conversion of lawn to meadow and wildflower areas began in 1992, and additional acreage was converted each year. Since then, CIGNA has changed 169 acres of lawn to grassy meadows and 26 acres to annual and perennial wildflower fields (figure 50). Although the changes have received positive reviews from the press and many employees enjoy the enhanced views, there was some initial resis-

FIGURE 50. A late summer view of a wildflower meadow, formerly an Industrial Lawn, at CIGNA Corporation's corporate park in Bloomfield, Connecticut. Photo: Sally Atkins, 1999.

tance and mixed feelings associated with the changed landscape. Some employees missed the traditional manicured lawns, especially in winter, when the wildflowers were dormant. Nor did some employees appreciate the increased diversity of insects and mammals in the meadows or realize the importance of CIGNA's environmental goals.

Fortunately, Sanderson understood that change can be threatening and that resistance is natural. He focused his efforts on public relations and education, writing articles about the new lawn care regime for the internal company newsletter. During a companywide environmental awareness day, the grounds department maintained a booth to disseminate information about grounds maintenance. The landscape changes have affected not only the employees but also members of the local retirement community, who have shown interest in the changes and have visited the grounds.

In addition to decreasing the lawn acreage, an equipment upgrade and an overall change in management techniques for the remaining 100 acres of lawn reflect the new corporate image. The lawn area is mowed weekly, but grass cuttings are now left to decompose, returning nutrients to the soil. Two diesel rotary mowers have replaced the five gasoline reel mowers, and even though these machines still contribute to noise and air pollution, they require less manpower to mow the acreage in the same amount of time.

Another change in management techniques has focused on the amount and frequency of chemicals and water applied to the remaining lawns. Chemical applications have decreased with the introduction of Integrated Pest Management, or IPM. The traditional sprinkler system has been upgraded to a smarter one that automatically shuts off when it rains. This combination of a new irrigation system and more ecologically sound lawn care techniques saves CIGNA 275,000 gallons of water a day,[10] which in turn has allowed CIGNA to decrease the number of wells it uses from thirteen to two.[11]

To effect the changes, CIGNA spent $63,000 over five years, which included new lawn mowers ($30,000 after the sale of the old lawn mowers), sprinkler system upgrades and maintenance ($20,000), and wildflower plantings ($13,000).[12] Several hundred thousand dollars are saved annually from reduced application of pesticides, fertilizers, and water, in addition to lower equipment maintenance and labor costs. What used to make sense to CIGNA—taking pride in a perfect lawn—no longer makes sense. "What are you going to do, spend $5,000 on dandelion control?" says McDonald. "And, okay, you kill the dandelions and you have nice green grass. But what's happening to that chemical? We're thinking more and more about not just now but twenty years down the road."[13]

Sanderson believes that it is important for CIGNA, a health care–related corporation, to become a corporate leader in environmental landscape management. Because CIGNA is in the public eye, it can make many people aware that corporations should be concerned about the environment and be responsible community neighbors. The combination of environmental awareness and economic concerns at CIGNA has allowed for the acceptance of this bottom-up approach to environmental management. Although CIGNA's campus represents a small portion of America's lawns, its new philosophy and role as a model for other businesses are making a difference.

Other Corporate Cases of Wildlife Habitat Enhancement

Two Johnson & Johnson companies in New Jersey, Ortho Pharmaceutical Corporation and Ethicon, have converted some of their lawn acreage to wildflower fields.[14] In addition, prairies have been embraced as a landscape form by a number of companies in the Midwest, among them General Electric, Sears Roebuck & Company, and CUNA Mutual Assurance Company.

Since 1988 the nonprofit organization Wildlife Habitat Council (WHC) has worked with businesses to help them manage their undeveloped land as wildlife habitats. This work can involve restoration, protection, enhancement, or creation of suitable wildlife habitat and species management. WHC states that more than a million acres of corporate land has been converted to wildlife habitat. The partnership between the company and WHC is a win-win situation because it benefits the environment, corporate public relations, and employee morale. WHC provides a third-party credibility to more than 650 corporate wildlife sites and an objective project evaluation. To receive WHC certification, the projects must be at least a year old and fully documented.[15]

OWENS CORNING

There are many success stories of companies becoming more environmentally sound. The Owens Corning World Headquarters facility in Toledo, Ohio, is one of the corporate sites WHC has certified. As a provider of energy-efficient products, Owens Corning wanted to meet the challenge of being an important environmental leader. Its corporate environmental leadership asserts that it is moving the company beyond regulatory compliance and into proactive environmental action on several fronts. Regarding the management of the company's property, thirteen out of the twenty planted acres were designed as restored native prairies in 1996 by one of this book's authors, Diana Balmori (figure 51).

Ideally, after the third season of a prairie's reestablishment, an annual prescribed burn can be used to maintain the area instead of mowing. Fire stimulates the plants to grow better, promotes seed germination, and controls noxious plants that can choke out some of the less hardy native plants. If burning is prohibited, as it is at Owens Corning owing to its location in downtown Toledo, the prairie is mowed once or twice a year after the seeds have matured and dispersed. The annual cost per acre for the established prairie has been $140, about one-fiftieth the cost of $6,675 for turf (table 1). According to another study, a General Electric facility spent $1,500 per acre annually to maintain its twenty-three acres of lawn (total $34,500), but spent only $25 per acre per year (one-sixtieth the cost) to maintain its eighty acres of prairie landscape (total $2,000).[16]

THE SUBURBS: LAWNS AND THE HYDROLOGIC CYCLE

Water is our most precious natural resource. As human numbers and enterprises increase, access to water will become a greater factor in tensions within and between localities, regions, and even nations. In some places, water scarcity may be to

FIGURE 51. Owens Corning World Headquarters in downtown Toledo, Ohio, features a restored prairie. Photo: Kurt Warner/Creative Advantage. Reproduced with permission from Balmori Associates.

the twenty-first century what oil price shocks were to the 1970s, a major source of economic and political instability.[17]

Lest we think that we Americans are immune to water scarcities, recall that in earlier chapters we reported the heavy usage of irrigation water to support lawns in both arid and humid regions. Also consider the huge investments by western cities for the purchase and use of irrigation water previously used for agriculture. In the past decade, however, many communities have begun to pay attention to regulating lower-priority water usage, such as watering lawns. Clearly, there are ways of meeting scarcity by shifting priorities. There are also ways of increasing potable water supplies through a different approach to managing the water cycle other than the one currently used in most suburban-urban communities.

Every year, about 1.4 million single-family homes and condominiums are built in the United States.[18] Hundreds of thousands of acres of undeveloped and agri-

TABLE 1.

One-year cost comparisons between turf and prairie for
Owens Corning World Headquarters, Toledo, Ohio

	Turf	Prairie
	(costs for 5.25 acres)	(after establishment, costs for 13 acres)
mowing	$15,878	$1,815
fertilization	$4,550	—
aeration	$11,698	—
irrigation	$2,918	—
cost per acre	$6,675	$140

Source: Owens Corning Headquarters, 1999

cultural land and fields and forests are converted to suburban subdivisions surrounded by lawns. In the process, water quality and quantity are damaged, the
recharge of aquifers is limited, streams and floodplains are altered in structure and
function and even destroyed, biodiversity is lost, and the quality of the suburban
environment is compromised.

From Undeveloped to Developed Land

Let us consider how these negative impacts come about in the development of
many subdivisions. As a reference point, think of a hypothetical tract of humid
eastern forestland undergoing suburbanization. We wish to follow how the movement of water out of the forest ecosystem changes as forestland is converted to a
housing subdivision.

In our hypothetical forested ecosystem, during one year 45 percent of the rainwater falling on the forest is returned to the atmosphere as evapotranspiration—
that is, evaporation of water drawn from the soil and evaporated *through* leaves, plus
evaporation *from* surfaces, such as soil, rocks, and leaves. Ten percent runs over the
soil surface and leaves the ecosystem, and the remaining 45 percent becomes deep
seepage or rainwater that seeps through the soil and out of the bottom of the
ecosystem (figure 52). Deep seepage recharges groundwater storage sites, such as
aquifers, that are capable of storing large amounts of water. Groundwater harvested
through wells becomes a source of water for humans. Some groundwater, moving
slowly through the sub-ecosystem mass, emerges in springs and seeps, becoming
stream flow. During dry periods, stream flow is wholly dependent on groundwater.

As our forested area undergoes suburbanization, the natural water cycle is replaced

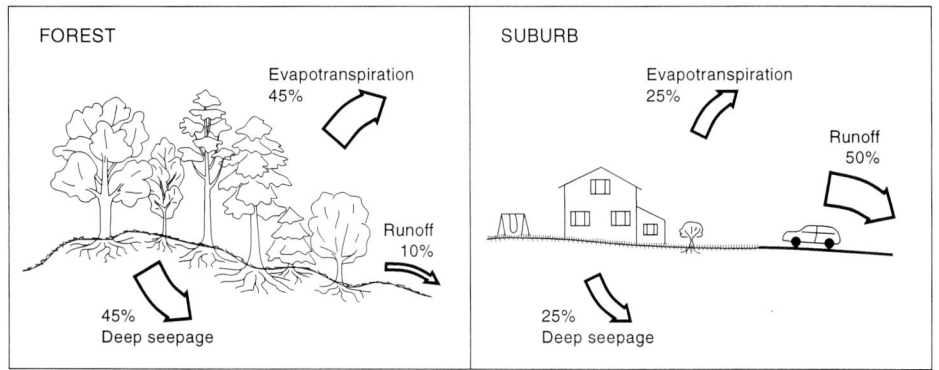

FIGURE 52. A comparison of the effect of conversion of an intact forest with a suburban house lot on the hydrologic cycle. Numbers are hypothetical. Adapted by Susan Hochgraf from a sketch by F. Herbert Bormann, 1999.

by what Robert Zimmerman, Jr., of the Charles River Watershed Association in Massachusetts calls a man-made water cycle.[19] In this cycle, rainwater falls on lawns, whose soils have been made less permeable to water by the bulldozing and reshaping of surfaces during construction and by compaction from heavy machinery (figure 53). The rainwater also falls on impervious surfaces, such as roads, driveways, sidewalks, parking lots, shopping malls, and rooftops, that are created in suburbanization (figure 54). Rather than penetrate into the ground, water from impervious surfaces collects and runs off, being added to the runoff from relatively impermeable surfaces like lawns. A typical feature of the traditional subdivision is that impervious and relatively impervious surfaces are connected to street drainage systems. These pipe systems are specifically designed to collect and remove storm water rapidly and empty the water into streams. As suburbanization proceeds, runoff increases dramatically over to undeveloped land, while evapotranspiration into the atmosphere and deep seepage to groundwater decreases substantially. Floods are more frequent, and there is an accelerated loss of drainage water out of the local area.

Streams, which nature engineered for more modest flows, now carry more water than they can handle. The problem is not just more water but the fact that runoff from suburban areas comes in a great rush after each rainstorm. Any observer can see this by observing a storm drain during, and shortly after, a heavy rain. Suburbanization also often channels streams or straightens their natural curves. This contributes to increased storm flow and the problems that ensue from it (figure 55).

Suburban storm flows have a higher velocity than flows from undeveloped land. The ability of storm flows to erode streambeds and to widen stream channels increases dramatically as the speed of flowing water increases. As the speed of storm

FIGURE 53. A golf course under construction, showing massive movement of earth and disturbance of the natural hydrologic cycle. Photo: F. Herbert Bormann.

FIGURE 54. A shopping mall where little effort has been made to increase water infiltration. Reproduced with permission from Yale School of Forestry and Environmental Studies, Center for Coastal and Watershed Systems.

FIGURE 55. *Top,* a small stream in a relatively natural condition; *bottom,* further downstream in a channelized condition. Reproduced with permission from Yale School of Forestry and Environmental Studies, Center for Coastal and Watershed Systems.

flow doubles, for example, erosive capacity increases four times; when storm flow increases four times, erosive capacity increases sixteen times.[20] Even modest rainfalls, in suburbanized areas with significant imperviousness, can produce storm flows that damage drainage streams. These streams become degraded as storm flows erode banks, deepen bottoms, and pollute streams with eroded sediments.

IMPERVIOUS SURFACES

Hydrologists have found a strong correlation between the health of a drainage stream and the extent of impervious surfaces in its drainage basin or watershed. As a general rule, Thomas Schueler, an expert on storm water management, estimates that with less than 10 percent imperviousness in a watershed the health of the stream can be considered protected (figure 56).[21] With increasing imperviousness, however, not only are banks and bottoms eroded but in-stream temperatures and nutrient loadings increase. This diminishes diversity in fish and aquatic species. Instead of being a general asset to the health, beauty, and diversity of the suburban landscape, the stream becomes an eyesore, often dirtied with trash (figure 57).

More runoff is not the only problem. Storm water, moving quickly across lawns, driveways, roads, and parking lots, picks up pollutants: pesticides, fertilizers, oil and grease, animal feces, solvents, metals from our disintegrating cars, plastic wrappers, and foam cups. The first flush (the initial water that runs off the surface) is the most polluted because the runoff picks up many of the substances deposited since the last rainstorm. In many suburbanized areas, storm water is a veritable pollution cocktail. Storm water and agricultural runoff are the primary causes of polluted rivers nationwide.[22] Clearly, how we manage the landscape is reflected in the degradation or health of our water resources. In Madison, Wisconsin, for example, investigators have identified lawns as a principal source of the phosphorus that has polluted Lake Mendota.[23]

As we reshape and pave over the land, sending more water to drainage streams, we are disconnecting rainwater from groundwater. In effect, the man-made water cycle reduces the rate at which groundwater supplies are recharged. Consequently, groundwater's contribution to stream flow is reduced, which can result in dry streambeds during dry periods. If the rate of pumping water from aquifers exceeds the rate of recharge by deep seepage from rainwater, then the water table will fall. More energy and money will have to be spent to pump water from deeper depths. In the long run, the aquifer will fail and new sources of water will need to be found.

ENVIRONMENTAL ZONING

Can we reverse the undesirable consequences of the suburban man-made water cycle by moving it back toward the natural water cycle of forests and fields? One such attempt is the Charles River Watershed Association's pioneering "environmental zoning" and its application to a cluster of towns in metropolitan Boston.

This type of zoning recognizes water as a limit to growth and attempts to answer questions like the one posed above. The association's plan addresses such issues as impervious surfaces, lawns, local renewable water supplies, water distribution systems, wastewater treatment, landscape management, groundwater recharge, runoff, and water quality. The primary goal of environmental zoning is to reduce runoff, increase deep seepage, and protect the quality and quantity of local drinking water. The valuable task the Charles River Watershed Association has undertaken makes one hope that it may soon tackle the restoration of Olmsted's Emerald Necklace, a hydrologic cycle green space project tied to the Muddy River's flow and ending in the Charles River.

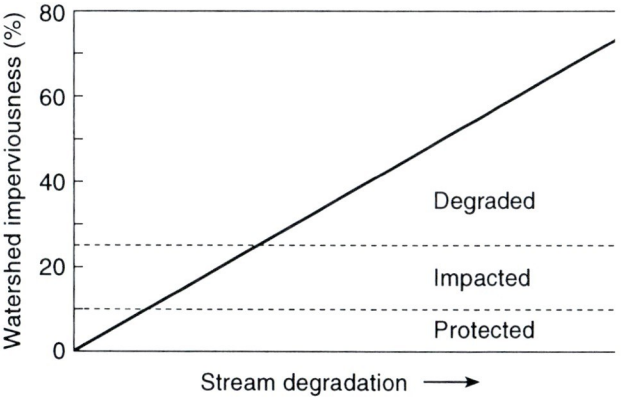

FIGURE 56. It is estimated that when 10–30 percent of a suburban watershed is covered by impervious surfaces, the health of the stream draining that watershed will be adversely affected, while at over 30 percent imperviousness, the watershed is considered degraded. Adapted from Thomas Schueler, Center for Watershed Protection, 1992. Reproduced with permission from Nonpoint Education for Municipal Officials (NEMO). Redrawn by Susan Hochgraf, 1999.

FIGURE 57. When streams contain trash, the area's water quality, aesthetics, and natural appeal are compromised. Reproduced with permission from Nonpoint Education for Municipal Officials (NEMO).

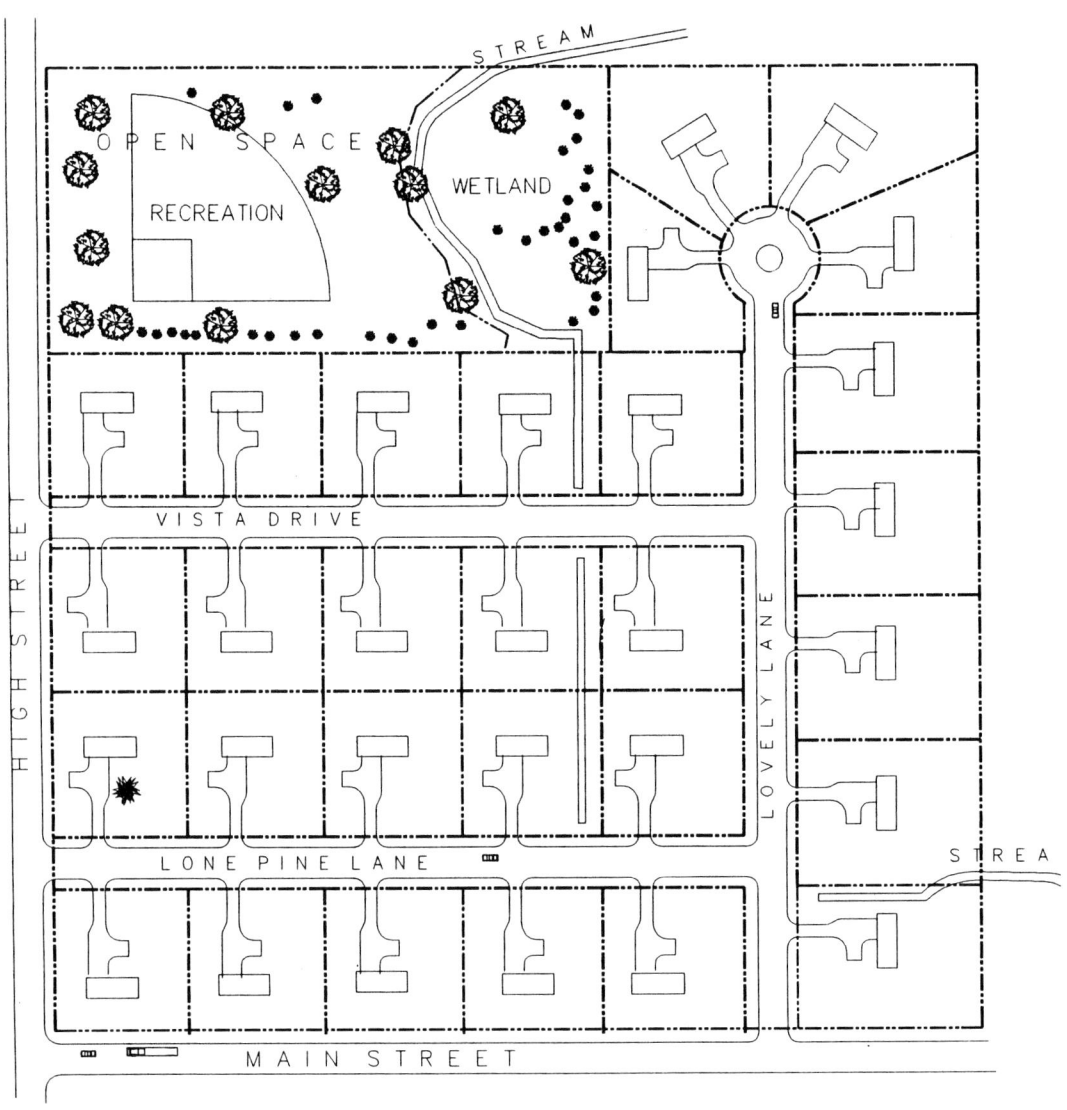

FIGURE 58. *Above,* a plan for a twenty-acre standard subdivision; *facing page,* a clustered subdivision on the same twenty acres. Clustering the same number of housing units substantially reduces impervious surfaces while increasing open space and buffer areas. Reproduced with permission from John Alexopoulos, University of Connecticut at Storrs.

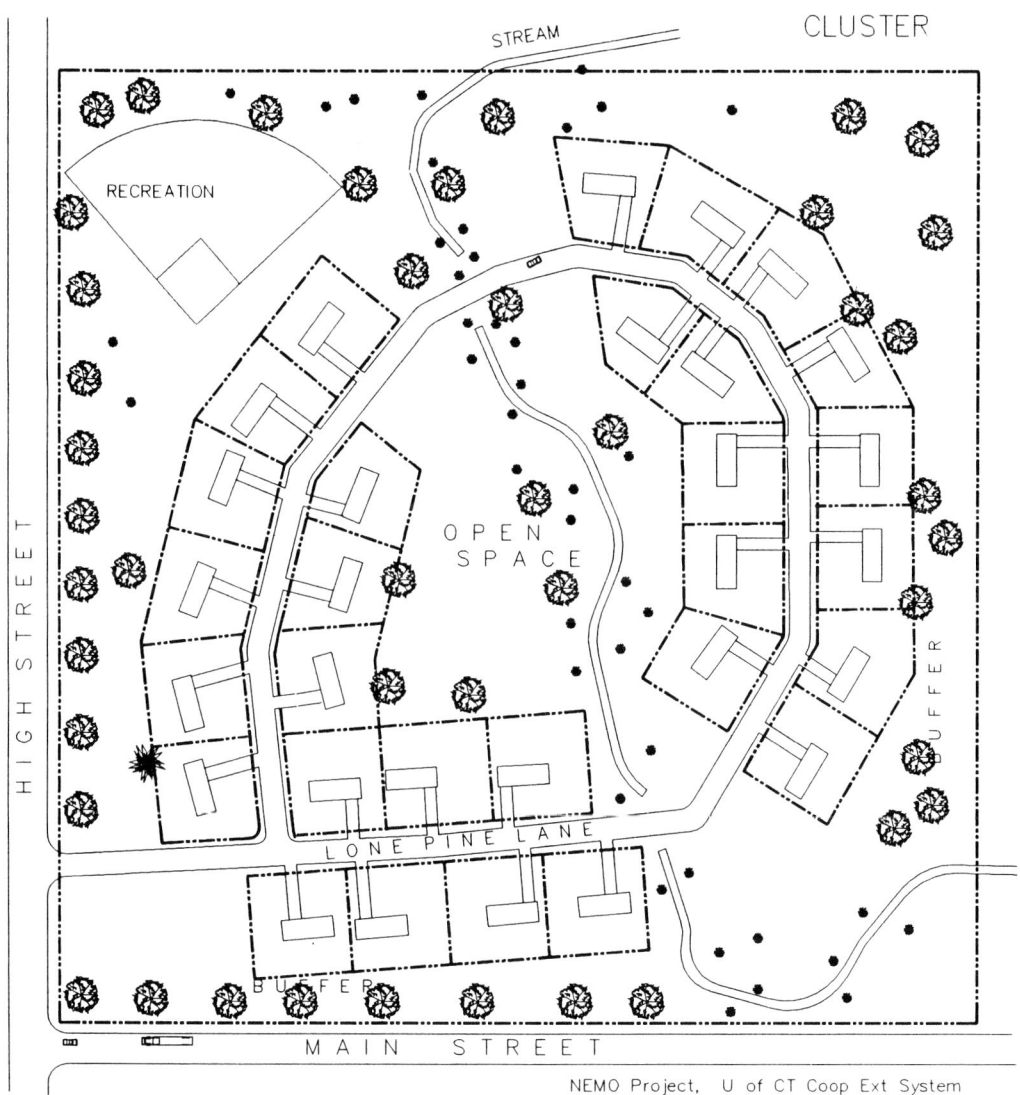

STREAM

RECREATION

HIGH STREET

OPEN
SPACE

BUFFER

LONE PINE LANE

BUFFER

MAIN STREET

NEMO Project, U of CT Coop Ext System

To best understand a community's water supply and wastewater problems, a systems approach, such as the association's environmental zoning, is necessary. Here we limit our discussion to the suburban landscape (lawns, fields, forests, recreation areas) and how, through landscape management, we can move the man-made water cycle toward the natural water cycle.

A basic tenet for improving the water cycle, during the early stages of suburbanization, is to retain intact as much as possible of the natural landscape. This will mean the retention of at least part of the natural water cycle's desirable features.

An example of how to preserve the natural environment during development is shown in figure 58. In this example, through cluster housing and smaller lot size, 50 percent of the twenty-acre area undergoing development would remain open space, preferably intact natural area, as opposed to 13 percent in a standard subdivision. Additionally, in the cluster subdivision, impervious surfaces would be reduced by 61 percent compared to the standard subdivision.[24] Runoff from impervious surfaces would be substantially reduced. If, during construction, care were taken to leave the areas destined for open space intact, then predisturbance rates of evapotranspiration, deep seepage, and runoff would be maintained in the open space. This would create a better balance of these rates for the developed area as a whole.

In the past fifteen years, many suburbs have been developed following a design principle called the New Urbanism. Essential to planning under New Urbanism is the establishment of public space, a pedestrian scale, and a neighborhood identity. Density is critical: the distance from the center to the edge of these new towns should be no more than a quarter of a mile or a five-minute walk. New Urbanism represents a reaction against sprawl and the concomitant view of land as an expendable commodity, though its reasons for planning a greater concentration of buildings stem more from a desire to reestablish community than for the sake of the land itself. This concept is a great idea, but it needs some additions. It would be helpful if New Urbanism set up guidelines on imperviousness and would use public space or greens to enhance the recharge of groundwater. Maintenance of the hydrologic cycle should be a principal objective.

As America continues to suburbanize, if care were taken to preserve as much as possible the hydrologic features of the natural landscapes being suburbanized, the ills associated with the man-made water cycle would be greatly diminished. If homeowners in a new development were to choose landscape designs that promoted greater seepage of water into the soil than Industrial Lawns, the area would have even less impact on the regional hydrologic cycle.

Where development has already occurred, many retrofitting steps can be taken to reduce the negative impacts of suburbanization on water quality and the hydrologic cycle. For example, as we have said many times in this book, minimizing the use of fertilizers, herbicides, and insecticides on Industrial Lawns would reduce the lawn's contribution to the pollution load of suburban storm flows. Converting Industrial Lawns to chemically free Freedom Lawns, meadows, or prairies would reduce suburban storm flow pollution even more. Infiltration of rainwater into the soil would be increased, particularly under meadows and prairies, because of deeper root growth and restoration of micro- and macro-soil organisms released from

TABLE 2.

Theoretical responses by landscape alternatives

	Industrial Lawn	Freedom Lawn	Meadow
Management			
mowing	regular	occasional	once/twice a year
fertilizer	yes	no	no
pesticide	yes	no	no
irrigation	yes or no	no	no
Ecosystem			
structure	one layer	one layer	more complex
plant diversity	monoculture	diverse	most diverse
soil organism diversity	narrow, adapted to pesticides and fertilizers	broader	broadest
infiltration of water	modest	modest	most
bird diversity	modest	more	most
Hydrology			
use of city water	yes or no	no	no
quality of runoff	low	good	best
groundwater recharge	modest	modest	best
storm flow contribution	modest	modest	least
Associated social costs			
fossil fuel use	most	much	least
landfill contribution	most	least	least

Note: Illustration adapted by Susan Hochgraf from a sketch by F. Herbert Bormann, 1999.

chemical control. This, in turn, would reduce runoff and increase groundwater recharge (table 2).

Collectively, through landscape design and maintenance, homeowners can do a great deal to reduce imperviousness and pollution. Over time, suburbanites could enjoy an improved quality and quantity of local water supplies while saving money and improving their aesthetic experience.

BAYSCAPES

The Chesapeake Bay, one of the largest and most productive estuaries in the world, is one of America's most troubled estuaries. Caring for its waters entails an environmental commitment from the millions of people who live within its 64,000-square-mile watershed.

In 1994, the Alliance for the Chesapeake Bay and the U.S. Fish and Wildlife Service developed a program known as BayScapes to encourage an improved environment for the region. The goal is to protect soil and water resources within the watershed while simultaneously lessening the pollution that finds its way into the bay.

The program promotes environmentally sound landscapes that reduce or prevent pollution and enhance wildlife habitat. Homeowners are taught to use fewer chemicals, employ naturalized vegetation, conserve water, reduce the use of small gas-powered engines, create wildlife habitat, and enhance biodiversity. BayScapes encourages homeowners to become responsible natural resource managers.

RAIN GARDENS

Planners in the mid-Atlantic region are advocating that a new landscape element, called Rain Gardens or bioretention areas, be incorporated into new subdivisions to achieve better water quality and to increase groundwater recharge. The principle underlying Rain Gardens is that suburban landscapes should be designed to funnel storm flow from lawns and impervious surfaces into many naturally occurring or constructed low-lying areas. There, some of the pollutants will be filtered out as the storm flow sinks into the ground, improving water quality. Not only does this infiltrating water help to recharge groundwater, but also it reduces storm flows and their damaging effects on drainage streams.[25]

HIGHWAY RIGHTS-OF-WAY

Four million miles of highways slice through and pave over the American landscape. These arteries, also known as turnpikes, thoroughfares, parkways, freeways, expressways, and superhighways, link flows of people, goods, and services vital to the health, welfare, and growth of the nation. This system carries serious environmental costs, which fortunately can be mitigated with thoughtful highway designs and management.

An important environmental effect of highways is on the hydrologic cycle and

FIGURE 59. Millions of acres of lawns line American highways. Reproduced with permission from Yale School of Forestry and Environmental Studies, Center for Coastal and Watershed Systems.

the problems associated with the man-made water cycle. Many of these concerns are related to imperviousness and involve engineering solutions that are beyond the scope of this book. Other problems are related to the design and management of highway rights-of-way, a topic intimately involved with lawns and central to our book.

A major issue is the highways' impact on aesthetics and biodiversity. There is a growing recognition that aesthetic and environmentally sound landscape alternatives exist for highway departments' love affair with the well-manicured lawn (figure 59). Yet demonstrating that a Freedom Lawn or a meadow (figure 60) is not a neglectful method of roadside management has been difficult in the face of skeptics, whose attitudes are based on the culturally ingrained concept of the Industrial Lawn.

The Texas Department of Transportation began a highway beautification program in the 1930s. Wildflowers were initially planted for control of soil erosion and later for aesthetics, but the program gained national attention when, inspired by Lady Bird Johnson, two Texas congressmen sponsored legislation, passed in 1984, that required states to use 0.25 percent of federal highway funds for planting wildflowers along federal highways.[26] This may seem like a negligible amount, but considering the billions of dollars spent on federal highways each year, the requirement

FIGURE 60. Self-sustaining wildflower displays along a Texas highway. On this particular Austin right-of-way, springtime bluebonnets are succeeded in summertime by coreopsis and Mexican hat. Photo: Jack Lewis/Texas Department of Transportation.

assigns millions of dollars each year for wildflower patches while providing a strong stimulus for states to search for aesthetically pleasing, environmentally sound, and economically viable alternatives to the well-manicured roadside lawn.

Wildflower Patches, Meadows, and Prairies

One option, now found in many states, is the wildflower patch or bed. Highway departments employ agricultural techniques (such as seedbed preparation, fertilizing, seed sowing, and weeding) to create a flower patch.[27] Because it takes two to three seasons for the perennials to reach their potential, annuals are added to provide a splash of color in the interim. Thousands of these patches now grace rights-of-way throughout the country, which for the most part bring satisfaction and pleasure to motorists. North Carolina's highway beautification program began in 1985 with 12 acres of wildflower beds; now 3,500 acres of these beds (versus 300,000 acres of turf) are installed and maintained along roadsides throughout the state.[28] Many of these patches, however, require renewal every few years and thus bring relatively high long-term economic and ecological maintenance costs.

Meadows, by contrast, are low-maintenance areas of naturally occurring vegetation that consist of native, naturalized, and introduced species of grasses and other herbaceous wildflowers, common to many of the better-watered sections of the country. Although the terms *meadow* and *prairie* are sometimes used interchangeably, prairies are generally regarded as expressions of the original grasslands that covered the midsections of the continent when Europeans arrived, whereas meadows are mostly products of human management.[29]

Meadows and prairies may be thought of as Freedom Lawns that are mowed or burned once a year or so. These plant communities are often characterized by beautiful native wildflowers such as hawkweeds, daisies, and black-eyed Susans in the Northeast, bur-marigolds in North Carolina, sunflowers in Kansas, bluebonnets and Indian paintbrushes in Texas, and poppies in California.

Benefits of Alternative Landscapes

Meadows and prairies have substantial ecological, aesthetic, and economic benefits compared to the conventional, frequently mowed, and often herbicided turf rights-of-way and are an improvement over wildflower plantings. Meadows and prairies are nontoxic, ecologically diverse landscapes that are cost- and energy-effective, and they ought to be incorporated in initial roadway designs.

ECOLOGICAL GAINS

Biological diversity increases when existing rights-of-way turf is converted to prairie and meadow plant communities. They provide important grass-shrub habitat for a variety of insects and mammals, such as American kestrels, which feed on grasshoppers and mice, meadowlarks and quail, which nest in the undergrowth, and honeybees, migrating hummingbirds, and butterflies, which depend on wildflower nectar for food; these pollinators in turn ensure plant reproduction. When the population of eastern monarch butterflies migrates between Canada and central Mexico or the western monarch population migrates from Canada to California, milkweeds growing along roadsides are essential for the butterflies' survival. Mowing and herbicide applications are considerable threats to these butterflies.[30]

Increasing biodiversity where we can is especially important because the United States is in the midst of a biodiversity crisis. For reasons ranging from degradation and outright loss of ecosystems to invasions by exotic species, one-third of our native species are now at risk.[31] Development of meadow and prairie rights-of-way can provide new habitat for endangered species and conserve precious genetic resources.

As previously considered (see table 2), in contrast to turf, meadow and prairie rights-of-way promote groundwater recharge, evapotranspiration, and improved water quality while also reducing storm flow. They also help to clean the air that moves near the surface of the ground. Compared to turf, the numerous leaves and stems and greater biomass more effectively filter some pollutants, such as dust and lead, out of the air stream that emanates from vehicle exhaust and the road surface. Such pollutants tend to be held in the roadside soil rather than running off in drainage water and contaminating surface or groundwater supplies.[32]

The individual states can improve their air and water quality by implementing these environmentally friendly landscapes, as well as ensure that they receive a full share of federal highway support. Under the Clean Air Act, the federal government can withhold a state's federal highway funding if the state does not meet national ambient air quality standards. Although this threat is rarely implemented, it does provide an incentive for state action.

AESTHETIC STIMULATION

The monotony of driving along turf-lined highways can induce a certain hypnosis. Meadows provide extra color and structural dimensions to our viewscape and alleviate the discomfort of the large amounts of time we spend in our cars. Senator Lloyd Bentsen of Texas once said that motorists may be "uplifted by the unique contribution of wildflowers indigenous to that part of the country through which they are traveling."[33] Color and form stimulates our senses, and seasonal changes enhance our appreciation for the natural cycles in nature. One hopes, too, that such meadows or prairies provide relief from the many stresses that characterize modern life.

ECONOMIC CONSIDERATIONS

Management of rights-of-way turf is an expensive and complex proposition involving different types of roads, varied mowing techniques, and an assortment of herbicides and herbicide applications.[34]

A primary economic benefit associated with reduced intensity of management is reduced costs for mowing and herbicide applications. Savings can vary enormously depending on such variables as the area of the rights-of-way, how much management intensity is reduced (for example, fewer mowings or, if converted to meadow, mowing just once a year), and local climatic influences. Savings per acre, whatever they are, have the potential to accumulate rapidly simply because of the enormous amount of roadside acreage. The Texas Department of Transportation, for example, manages seventy-seven thousand miles of highways and more than a million acres of rights-of-way.[35]

What level of saving occurs when a state applies reduced rights-of-way management to its whole system? Unfortunately no statistics are available, but we can make some educated guesses. In Massachusetts, reducing five or six turf cuts to one cut is estimated to reduce costs from $330 per acre to $50, saving $280 per acre.[36] If such a reduction were applied to all of Massachusetts's three thousand acres of roadside turf, a savings of about $1 million annually would result. One might imagine that for the nation as a whole, converting one-fourth of roadside turf into chemically free meadow, mowed once per year, would save many hundreds of million dollars.

A number of states have used their new rights-of-way to generate revenues. Some have designed wildflower routes to generate tourism dollars. Other states or communities have developed cultural events that incorporate wildflowers, prairie vegetation, and historical happenings. Several state commerce departments, agricultural extension services, private organizations and citizens maintain websites that alert tourists to local areas that are in bloom, while other websites offer wildflower photo tours.[37]

Nevertheless, even the arcane accounting practices of corporate America are not sufficient to consider the value of associated environmental savings, such as higher water quality, fuller aquifers, reduced pesticide levels, less contamination of food chains, greater species diversity, improved human health, and lower contribution to such problems as global climate change.

LAWNS AND BIRD MIGRATION

The twice yearly migrations of millions of birds such as robins, kingfishers, waterfowl, hummingbirds, vultures, hawks, and warblers are extraordinary natural events. Although an aspect of humans' fascination with avian migration is the biological navigation system that underlies the birds' ability to make these long voyages, we need to be aware of the ecological importance of these billions of birds in both their summer and winter residences.

Wherever they are, birds perform such vital ecosystem functions as seed dispersal, predation of insects and rodents, pollination, and partial regulation of food chains. Relative success during migration can thus affect the ecological health of large areas. During migration, birds need adequate food, water, and habitats that provide shelter and a place to rest. Land development and habitat alteration occurs in both northern and southern hemispheres, which makes it important to consider the essentials for migratory success in both public and private development plans.

Activities in our backyard not only can defeat our desire for clean air and water but also can affect significant global biological phenomena such as bird migration. For the most part, Industrial Lawns provide little benefit for most migratory birds: food is limited, shelter is almost nonexistent, deleterious chemicals are applied repeatedly, and machines make regular incursions (see table 2).

By contrast, yards landscaped with more complex ecosystems, like Freedom Lawns and meadows, provide migratory birds with a relative diversity of habitats and food sources, such as grains, berries, and a large variety of invertebrates that feed or rest on plants. Not only are these food sources available during fall migration, but some plant and animal foods overwinter and are available for returning spring migrants.

Migration Routes

The diversity of bird species that travel a variety of distances on numerous routes, or flyways, between their wintering and breeding grounds is staggering. Billions of songbirds, raptors, shorebirds, and waterfowl that breed in North America have wintering grounds south of the United States border. Out of 9,672 species of birds worldwide, at least 660 species nest in North America, and more than two-thirds of these, known as Neotropical migrants, travel seasonally outside the United States.[38]

The term *migration route* describes a species' approximate movements between its breeding and wintering grounds, but it does not depict an individual bird's exact course. It was once believed, based on migratory waterfowl banding studies from the 1930s, that four flyways across the United States (Atlantic, Mississippi, Central, and Pacific) were the definite routes for all birds. It is now known that many bird species migrate along variable and widely dispersed routes, although more or less regular migratory paths can be identified for various species (figure 61).[39]

The National Audubon Society and the Cornell Laboratory of Ornithology support a website (www.birdsource.org) through which citizens can participate in bird counts and help gather research data. Especially exciting is the possibility of bird lovers' being able to track bird migration by radar on the computer and report sightings in their backyards and local parks. This site also has a detailed and informative section that educates backyard birders about the hazards to birds of applying many commonly available pesticides and herbicides to their lawns.

Decline of Bird Species

A number of migratory Neotropical species are in decline but not yet in crisis. If trends continue, however, some researchers fear significant losses of migrating species or a serious decline in the number of individuals making up a species' population. This could alter the structure and function of ecosystems over large areas. If, for example, bird species that are the major dispersers of a particular plant species become extinct, the plant may not be able to colonize new areas. In time, the plant could be diminished as a significant member of the ecosystem. And in turn, a host of organisms that are directly or indirectly dependent on that plant for their livelihood could undergo decline. Clearly this is a highly speculative scenario, but it illustrates how interconnected life is and how the loss of a migratory bird species could cascade through a regional ecosystem.

Some reasons for the decline in bird species richness are habitat destruction and alteration, environmental contaminants, and competition for resources along migratory routes as well as in summer and winter grounds.[40] Along the various migratory routes, birds are often squeezed through bottlenecks where they are particularly vulnerable to landscape manipulations, such as suburban yards dominated by the Industrial Lawn.

On the way southward toward winter destinations, migratory birds often rely on key, widely scattered stopover points that provide water and abundant food resources. Vast concentrations of birds depend on these refueling sites to replenish their fat reserves before undertaking long, nonstop flights over land or water. Sometimes more than 80 percent of the North American population of some species

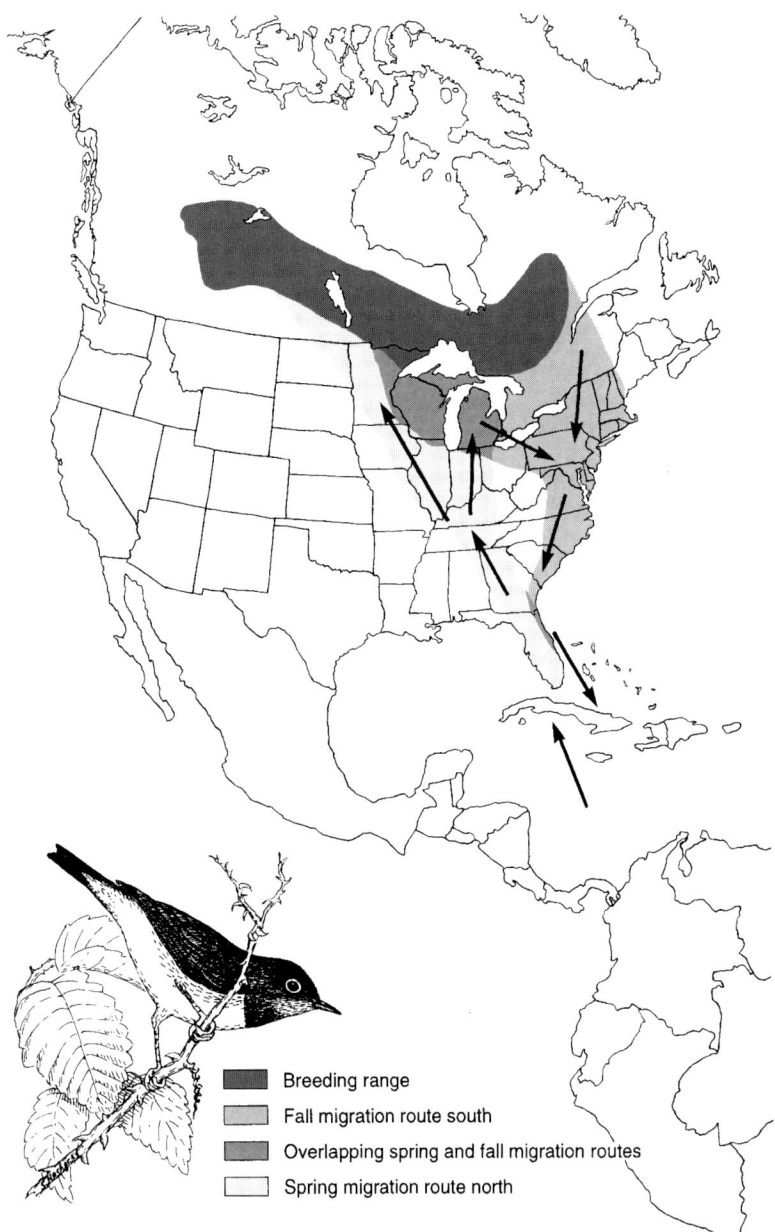

FIGURE 61. Breeding range and migratory routes of the Connecticut warbler. From its breeding range in the northern United States and southern Canada, the Connecticut warbler migrates east in the fall to New England, south along the Atlantic coast to Florida, and across the West Indies to winter in South America. In the spring the warbler returns by migrating northwest across the Allegheny Mountains and the Mississippi Valley. Reproduced with permission from the U.S. Department of the Interior, Fish and Wildlife Service. Illustration by Bob Hines redrawn by Susan Hochgraf, 1999.

Breeding range
Fall migration route south
Overlapping spring and fall migration routes
Spring migration route north

congregates at one of these stopover sites.[41] When a key stopover area is developed for residential or commercial purposes, with only humans' needs in mind, the result can be catastrophic not only for these migratory birds but for local birds and the whole environment.

Critical spots where a variety of migrating birds concentrate have been identified worldwide. Five such critical sites in North America that annually support millions of migrating shorebirds are Washington's Gray's Harbor, Kansas's Cheyenne Bottoms, Alaska's Copper River Delta, eastern Canada's Bay of Fundy, and the land surrounding Delaware Bay in New Jersey and Delaware.[42] Throughout these areas, birds use whatever habitat is available.

CAPE MAY, NEW JERSEY

Located at the southern tip of New Jersey, Cape May is a key component of the Atlantic coast flyway. The peninsula, which consists of beaches, wetlands, and wooded and developed land, is a critical stopover site for millions of Neotropical migrants and birds of prey. Incredibly, more than two hundred bird species were recorded on one day during New Jersey Audubon's world series of birding in Cape May County, including such rarer species as peregrine falcons, sharp-shinned hawks, Cape May warblers, bobolinks, and American woodcocks.[43] The migrations and other features of bird life are major attractions to the area and as a premier birding spot, bird migration is economically important. It is estimated that tourists put $31 million into the local economy in 1998.[44]

In the early 1990s, the New Jersey Division of Fish, Game, and Wildlife's Endangered and Nongame Species Program (ENSP) analyzed twenty years of habitat change and confirmed a rapid loss of migratory bird habitat (figures 62 and 63); this was a threat to the cape's natural heritage and to its commercial well-being. As a result, the department instigated the Cape May Stopover Protection Project, which featured an innovative, integrated approach to migrant land bird habitat conservation: the project engaged willing landowners, municipal and county planners, and open space land managers in voluntary partnerships to reverse the trend of habitat loss in the southernmost several miles of Cape May County.[45] As the coordinating agency, the ENSP oversaw all aspects of the project and was responsible for delineating critical migratory bird habitat, identifying protected areas, and disseminating information to land planners and managers.[46]

The Nature Conservancy focused on relatively large parcels of land and encouraged landowners such as developers, campground owners, farmers, churches, and private residences to manage their land in ways friendly to wildlife. Developers were urged, for example, to cluster homes and include forested areas in new

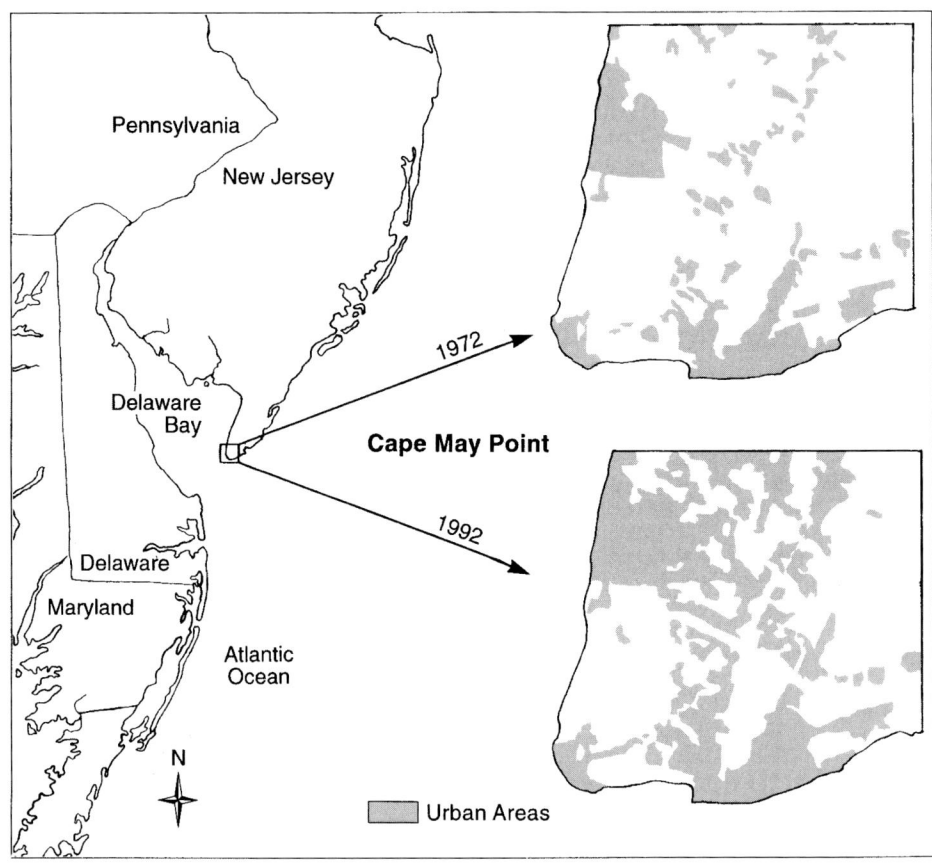

FIGURE 62. The New Jersey Division of Fish, Game, and Wildlife's habitat analysis for the several miles of the Cape May peninsula. In two decades suburbanization doubled while suitable habitat for migratory birds decreased by 40 percent. The study served as the basis for new land management practices that would provide high-quality migratory bird habitat. Reproduced with permission from the State of New Jersey Division of Fish, Game, and Wildlife's Endangered and Nongame Species Program. Redrawn by Susan Hochgraf, 1999.

subdivisions. Ultimately, the conservancy assisted each landowner in developing a site-specific, long-term habitat conservation plan.

The Stopover Project included new landscaping ordinances. With people continuing to flock to the area for new housing, in the spring of 1999 the towns of West Cape May and Cape May created their own landscape and vegetation standards by adapting one from nearby Cape May Point.[47]

The stopover program also encourages small homeowners in the vicinity of Cape May Point to plant native or naturalized vegetation to provide food for birds at the right time of the year (figure 64). To help guide citizens, New Jersey's Conserve

FIGURE 63. Stone Harbor, a town on the Cape May peninsula, where homes with stone yards have replaced a substantial portion of the natural vegetation. Photo: Steve Eisenhauer.

Wildlife Foundation secured funds to create a model backyard habitat at the Cape May Bird Observatory's Center for Research. The habitat's designer, Karen Williams, says that the garden is designed to show visitors a variety of ways that a yard can provide for wildlife. The area features butterfly gardens, tree and shrub borders, a meadow, and a pond. Its aim is to empower people to develop a style of gardening that can help wildlife, reduce the need for supplemental water and polluting chemicals, reduce overall gardening costs, and be beautiful. Williams believes that alternative landscaping is no more costly than traditional landscaping.[48]

The New Jersey Audubon Society, ENSP, and Conserve Wildlife Foundation provided workshops for 300 to 400 homeowners and followed up with site visits and specific recommendations for about 120 of the participants. Homeowners who successfully relandscaped had their yards certified as wildlife sanctuaries. ENSP also secured a grant from the Environmental Protection Agency, which provided workshop participants with a one-hundred-dollar voucher for appropriate plants at local nurseries.

The workshops were a huge success; they created a network of community educators. Several attendees who were business owners encouraged their clients to

This property has been designated a

WILDLIFE SANCTUARY

for migratory birds, butterflies and other wildlife,
by providing plants for food and cover

as part of the
Cape May Stopover Protection Project,
For further information write:
NJ Endangered and Non game Species Program
Division of Fish, Game and Wildlife
2201 County Rt. 631, Woodbine, NJ 08270

VEGETATION STANDARDS

Commonly, municipal ordinances mandate specific landscaping requirements for new developments. However, less common are communities that support strict standards to maintain vegetation. One such community that has a progressive landscaping and vegetation plan is the Borough of Cape May Point, New Jersey.

Historically, the residents of Cape May Point and the surrounding towns believed that stone yards, where native vegetation was excluded and herbicides were used to control weeds, were the best "low-maintenance" landscapes. The economic boom of the early 1980s brought major development schemes to this coastal area, which mainly consists of sand dune habitats. Community residents became concerned with the environmental degradation and habitat loss for migratory birds that accompanied this rapid development. In 1988, according to Anita van Heeswyk, head of the Cape May Point Environmental Commissions Board, the community passed an ordinance that recognized the multiple benefits of diverse vegetation: it filters storm water, acts as a windbreak, contributes shade, increases property values, and provides habitat for migratory birds.

Part of Cape May Point's original landscaping and vegetation plan states that "all driveways and parking areas shall be a pervious surface. . . . Plant two (2) trees for each tree removed [outside a building footprint], . . . such replacement trees shall be of at least two (2) inches in diameter. . . . [The property] shall retain as much of the natural vegetation as is possible." In 1998, Cape May Point amended the plan, now mandating that at least 60 percent, rather than at least 50 percent, of the overall residential lot must be covered by vegetation. If a house takes up 30 percent of the lot, therefore, only the remaining 10 percent area may be used for driveways, walkways, and patios.

FIGURE 64. *Above,* a grass-covered yard in 1981 with few provisions for wildlife (photo taken from across the street); *right,* the front yard ten years later (looking from the house toward the street). Note the lamppost in both photos. Steve Grout has relandscaped his property to provide food, shelter, and water for a variety of butterflies, small mammals, and migratory birds (among them, American redstarts, black-and-white warblers, common yellowthroats, and sharp-shinned hawks). Phlox, sedum, verbena, and milkweed have been planted for butterflies. Japanese black pine provides shelter for birds throughout the year, while bayberry, staghorn sumac, red mulberry, and Virginia creeper provide seasonal food supplies. Photos: Stephen Grout.

FIGURE 65. In Cape May, New Jersey, Ro Wilson maintains a mowed path around her meadow, which permits the close observation of nature. Photo: Ro Wilson.

attend the workshops and informed them about alternative lawn care options. The workshops also served as a model for another New Jersey town to work with its community landowners.

Over the years, Williams has seen that it takes time for people to accept the need to change landscape management. She stresses that landowners can maintain mowed areas, but if they adopt a more relaxed approach to yield a mixture of grasses and weeds—a Freedom Lawn—then the essentials for wildlife are provided throughout the year. She accepts that not everyone will be won over by this alternative approach to land management. But community education is an important way to help change people's attitudes. Education can be as simple as erecting a sign in front of one's meadow stating that the area supports butterflies and migratory birds. Or a path can be mowed through a meadow so that the habitat can be observed and enjoyed at close range (figure 65).

AMERICA THE BEAUTIFUL

The two-hundred-year-old home of America's president, the White House, is visited annually by more than a million people.[49] Multitudes pass by it every day. Nearly every American has seen the White House on television, in newspapers and magazines, or on postcards. Not only does the White House stand as a symbol of democratic ideals and freedom, but by its very existence and maintenance it sends a message to the world of America's environmental values.

The nation's president could take environmental leadership farther by setting an example at the White House. Much as Thomas Jefferson created an ideal symbolic landscape at Monticello for his own time—a productive and beautiful farm for an earlier nation of farmers—the occupants of the first family's residence could create a landscape for our time.

With the initiation of a program called Greening the White House, important steps were taken in 1993 to reduce chemical use, water consumption, solid waste, and greenhouse emissions. By the mid-nineties the White House had adopted a lawn care regime that not only cost less but reduced consumption of natural resources.[50]

In the spirit of Thomas Jefferson, the White House could now put forth an image for the new millennium. Following the principles we have outlined in this book, the grounds should be a Freedom Lawn, a meadow, or a landscape of flowers of many species and seasons—a landscape not dominated by Industrial Lawn and not dependent on a continuous flow of water, fertilizers, and insecticides. The White House should be surrounded by a landscape that strives to conserve natural resources and maintain the natural hydrologic cycle, one that can flourish with minimal additions; a beautiful landscape shaped and designed with the best of what nature fosters in our country; a landscape that can stand as an example for the world.

Notes

CHAPTER 1: LOVE OF THE LAWN

1 Walt Whitman, *Leaves of Grass*, 1855, in *Complete Poetry and Collected Prose* (New York: Viking, 1982), 31.

2 J. Falk, telephone interview conducted by B. Milton, February 1992.

3 Tony Hiss, *The Experience of Place* (New York: Knopf, 1990).

4 Worthington Chauncey Ford et al., eds., *Journals of the Continental Congress, 1774–1789*, 34 vols. (Washington, D.C.: U.S. Government Printing Office, 1904–37), 30:230–31.

5 John Adams, letter to John Sullivan, May 27, 1776, in *Papers of John Adams*, vol. 4: *February–August 1776*, ed. Robert J. Taylor (Cambridge: Harvard University Press, 1979), 210.

6 Weyerhaeuser Company, *The Value of Landscaping*, vol. 4 of *Ideas for Today* (Tacoma, Wash.: Weyerhaeuser Nursery Products Division, 1986).

7 For a general reference on medieval gardens, see John Harvey, *Mediaeval Gardens* (Beaverton, Oreg.: Timber Press, 1981).

8 General references on French gardens include Elizabeth B. MacDougall and F. Hamilton Hazlehurst, eds., *The French Formal Garden* (Washington, D.C.: Dumbarton Oaks Research Library and Collection/Trustees for Harvard University, 1974), and William Howard Adams, *The French Garden, 1500–1800* (New York: Braziller, 1979).

9 For a general reference on English gardens, see David Watkin, *The English Vision: The Picturesque in Architecture, Landscape, and Garden Design* (New York: Harper and Row, 1982).

10 Simon Pugh, *Garden, Nature, Language* (Manchester: Manchester University Press, 1988).

11 For more information on William Kent, see Michael I. Wilson, *William Kent: Architect, Designer, Painter, Gardener, 1685–1748* (London: Routledge and Kegan Paul, 1984), and John Dixon Hunt, *William Kent: Landscape Garden Designer: An Assessment and Catalogue of His Designs* (London: A. Zwemmer, 1987).

12 Horace Walpole, *The History of the Modern Taste in Gardening / Journals of Visits to Country Seats* (New York: Garland, 1982), 264.

13 George B. Tobey, Jr., *A History of Landscape Architecture: The Relationship of People to Environment* (New York: American Elsevier, 1973).

14 David Watkin, *The English Vision* (New York: Harper and Row, 1982), 181–82.

15 Roger Turner, *Capability Brown and the Eighteenth-Century English Landscape* (New York: Rizzoli, 1985).

16 Yves Abrioux, *Ian Hamilton Finlay: A Visual Primer* (Edinburgh: Reaktion Books, 1985), 38.

17 William Cronon, *Changes in the Land: Indians, Colonists and the Ecology of New England* (New York: Hill and Wang, 1983).

18 Thomas Jefferson, *A Tour to Some of the Gardens of England / Travel Journals* (reprinted from *The Writings of Thomas Jefferson* [Washington, D.C.: U.S. Department of State, 1853–54, vol. 9], in *Thomas Jefferson: Writings* [New York: Library of America, 1984]), 627.

19 American Society of Landscape Architects, *Colonial Gardens: The Landscape Architecture of George Washington's Time* (Washington, D.C.: George Washington Bicentennial Commission, 1932).

20 Kenneth B. Jackson, *The Crabgrass Frontier: The Suburbanization of the United States* (New York: Oxford University Press, 1985).

21 John Brinckerhoff Jackson, "The Public Landscape," in *Landscapes: Selected Writings of J. B. Jackson,* ed. Ervin H. Zube (Amherst: University of Massachusetts Press, 1970), 157–58.

22 Jackson, *Crabgrass Frontier,* 60.

23 Jane Loudon, *Ladies' Companion to the Flower-Garden* (London: Bradbury and Evans, 1841), 207.

24 Samuel Orchart Beeton, *Beeton's Dictionary of Everyday Gardening* (London: Ward, Lock, 1874), cited in Christopher Thacker, *History of Gardens* (Berkeley: University of California Press, 1979), 233.

25 Andrew Jackson Downing, *A Treatise on the Theory and Practice of Landscape Gardening, Adapted to North America* (London: George Putnam, 1841; Washington, D.C.: Dumbarton Oaks Research Library and Collection/Trustees for Harvard University, 1991).

26 M. Christine Klim Doell, *Gardens of the Gilded Age: Nineteenth-Century Gardens and Homegrounds of New York State* (Syracuse, N.Y.: Syracuse University Press, 1986), 6.

27 Jackson, *Crabgrass Frontier,* 58.

28 Jackson, *Crabgrass Frontier,* 59.

29 Jackson, *Crabgrass Frontier,* 59.

30 Bruce Kelly, "Art of the Olmsted Landscape," in *Art of the Olmsted Landscape,* ed. Bruce Kelly, Gail Travis Guillet, and Mary Ellen W. Hern (New York: New York City Landmarks Preservation Commission and the Arts, 1981).

31 For a general reference on Alexander Jackson Davis, see Jackson, *Crabgrass Frontier,* and Jeffrey Kastner, "Alexander Jackson Davis," in *Three Centuries of Notable American Architects,* ed. Joseph J. Thorndike, Jr. (New York: American Heritage, 1981).

32 R. Cheatle, P. Farrant, and I. Latham, "Riverside, 1869," in *The Anglo-American Suburb,* ed. Robert A. M. Stern and John Montague Massengale (New York: St. Martin's, 1981), 24.

33 William H. Whyte, *City: Rediscovering the Center* (New York: Doubleday, 1988), 123.

CHAPTER 2: QUESTIONING THE LAWN

1 "Weed All About It," *People Weekly* 30 (1988): 45; Walter Stewart, telephone interview conducted by J. H. Connolly, February 1991.

2 Michael Pollan, *Second Nature: A Gardener's Education* (New York: Atlantic Monthly Press, 1991), 63–64.

3 Murray Blum, letter dated Apr. 23, 1992.

4 Joel Meisel, telephone interview conducted by J. H. Connolly, February 1991.

5 Charles Darwin, *The Origin of Species* (London: W. Clowes and Sons, 1859).

6 Henry David Thoreau, "Walking," in *Excursions* (Boston: Ticknor and Fields, 1863), 174, 188–190.

7 George Perkins Marsh, *Man and Nature; or Physical Geography as Modified by Human Action* (New York: Charles Scribner, 1864), 36.

8 William Vogt, *Road to Survival* (New York: William Sloane Associates, 1948).

9 Aldo Leopold, *A Sand County Almanac* (New York: Oxford University Press, 1949).

10 As cited in Robert C. Paehlke, *Environmentalism and the Future of Progressive Politics* (New Haven and London: Yale University Press, 1989), 18.

11 Rachel Carson, *Silent Spring* (Cambridge, Mass.: Riverside, 1962).

12 Carson, *Silent Spring,* 27.

13 Carson, *Silent Spring,* 173.

14 Carson, *Silent Spring,* 176.

15 Barry Commoner, *The Closing Circle—Nature, Man, and Technology* (New York: Alfred A. Knopf, 1971); Paul Ehrlich, *The Population Bomb: Population Control or Race to Oblivion* (New York: Ballantine Books, 1968).

16 Worldwatch Institute, *State of the World* (New York: W. W. Norton, 1984–2000).

17 Lester R. Brown, "The New World Order," in *State of the World 1991,* ed. Lester R. Brown et al. (New York: W. W. Norton, 1992), 5.

18 René Dubos, *Celebrations of Life* (New York: McGraw-Hill, 1981), 81.

CHAPTER 3: THE ECONOMIC JUGGERNAUT

1 National Gardening Association, *National Gardening Survey, 1998–1999,* Burlington, Vt.

2 Michael Pollan, *Second Nature: A Gardener's Education* (New York: Atlantic Monthly Press, 1991), 55–56.

3 U.S. Environmental Protection Agency, Office of Pesticide Programs, "Fact Sheet on Lawn Care Pesticide Use," 1996.

4 Augusta Goldin, *Grass: The Everything, Everywhere Plant* (New York: Thomas Nelson, 1977), 143.

5 F. William Ravlin and William H. Robinson, "Audience for Residential Turf Grass Pest Management Programs," *Bulletin of the Entomological Society of America* 31, no. 3 (1985): 45–50.

6 Professional Lawn Care Association of America and the Lawn Institute, "The ABC's of

Lawn and Turf Benefits" (Professional Lawn Care Association, Marietta, Ga., and Lawn Institute, Pleasant Hill, Tenn., n.d.).

7 "Sell the Environment to Increase Your Profits," *Lawn News* 7 (1991): 3.

8 *National Gardening Survey, 1998–1999.*

9 Goldin, *Grass.*

10 David Barboza, "Scientists Pursue Suburban Genetics in the Search for a Perfect Lawn," *New York Times,* July 9, 2000.

11 Whit Yelverton of the Fertilizer Institute, telephone interview conducted by B. Milton, February 1991.

12 For statistics on the fertilizer industry, see U.S. Department of Commerce, *United States Industrial Outlook '90* (Washington, D.C.: U.S. Government Printing Office, 1990).

13 R. Ringer of Ringer Corporation, telephone interview conducted by B. Milton, February 1991.

14 U.S. Environmental Protection Agency, Office of Pesticide Programs, *Pesticides Industry Sales and Usage: 1996 and 1997 Market Estimates,* by Arnold L. Aspelin and Arthur H. Grube, 733-R-99-001 (Washington, D.C.: U.S. Environmental Protection Agency, 1999), table 1.

15 Office of Pesticide Programs, *Pesticides Industry Sales and Usage,* table 2.

16 David Pimentel, "The Dimensions of the Pesticide Question," in *Ecology, Economics, Ethics: The Broken Circle,* ed. F. Herbert Bormann and Stephen R. Kellert (New Haven and London: Yale University Press, 1991), 59.

17 Cited by "The Dangers of Lawn Care," compiled by the editors of *Organic Gardening,* January 1989, 1.

18 "Lawn Care Industry Dilemma," *American Horticulturist* 69, no. 11 (1990): 4.

19 Turfgrass Council of North Carolina, *North Carolina Turfgrass Survey* (Raleigh, N.C.: North Carolina Crop and Livestock Reporting Service, 1987), 37.

20 U.S. Department of Commerce, U.S. Census Bureau, Economics and Statistics Administration, Manufacturing Industry Series, *U.S. Lawn and Garden Tractor and Home Lawn and Garden Equipment Manufacturing, 1997,* EC97M-3331B (issued October 1999).

21 U.S. Department of Commerce, *United States Industrial Outlook '90; National Gardening Survey, 1998–1999.*

22 U.S. Department of Commerce, *U.S. Lawn and Garden Tractor Manufacturing;* see also <www.doc.gov>.

23 Professional Lawn Care Association of America, <www.plcaa.org/press>, 2000.

24 U.S. General Accounting Office, *Lawn Care Pesticides,* 15.

CHAPTER 4: ENVIRONMENTAL COSTS

1 National Gardening Association, *National Gardening Survey, 1998–1999,* Burlington, Vt., 103.

2 City of Irvine, Community Development Department, *Sustainable Landscaping Guideline Manual,* 1991 Draft (Irvine, Calif.: Community Development Department, 1991).

3 Francesca Lyman, with Irving Mintzer, Kathleen Courrier, and James Mackenzie, *The Greenhouse Trap: What We Are Doing to the Atmosphere and How We Can Slow Global Warming,* World Resources Institute Guide to the Environment (Boston: Beacon, 1990), ix.

4 Jerry Martin, spokesperson for the California Air Resources Board, telephone interview conducted by Lisa Vernegaard, September 1992.

5 California Air Resources Board, *Technical Support Document.*

6 U.S. Environmental Protection Agency, *Nonroad Engine and Vehicle Emission Study,* Report No. 21A-2001 (Washington, D.C.: Environmental Protection Agency, Office of Air and Radiation, November 1991).

7 M. Talbot, "Ecological Lawn Care," *Mother Earth News* 123 (1990): 60–73.

8 Warren Schultz, "Natural Lawn Care," *Garbage* 2, no. 4 (1990): 26–34.

9 U.S. Senate, *The Use and Regulation of Lawn Care Chemicals: Hearing Before the Subcommittee on Toxic Substances, Environmental Oversight, Research and Development of the Committee on Environment and Public Works, U.S. Senate, March 28, 1990,* Senate Hearing 101-685 (Washington, D.C.: U.S. Government Printing Office, 1990), 1.

10 C. R. Frink, *Uses of Pesticides in Connecticut,* Connecticut Agriculture Experiment Station Bulletin No. 848 (Hamden: Connecticut Agricultural Experiment Station, 1987).

11 *National Gardening Survey, 1998–1999.*

12 O. P. Engelstad, ed., *Fertilizer Technology and Use,* 3d ed. (Madison, Wis.: Soil Science Society of America, 1985).

13 Warren Schultz, *The Chemical-Free Lawn* (Emmaus, Pa.: Rodale, 1989).

14 Samuel L. Tisdale, Werner L. Nelson, and James D. Beaton, *Soil Fertility and Fertilizers,* 5th ed. (New York: Macmillan, 1993). See also Lyman et al., *Greenhouse Trap,* and Engelstad, ed., *Fertilizer Technology and Use.*

15 Engelstad, ed., *Fertilizer Technology and Use;* Lyman, *Greenhouse Trap.*

16 U.S. Environmental Protection Agency, Office of Pesticide Programs, *Pesticide Industry Sales and Usage: 1996–1997 Market Estimates,* by Arnold L. Aspelin and Arthur H. Grube, 733-R-99-001 (Washington, D.C.: U.S. Environmental Protection Agency, 1999).

17 Henry F. Decker and Jane M. Decker, *Lawn Care: A Handbook for Professionals* (Englewood Cliffs, N.J.: Prentice Hall, 1988).

18 Frits van der Leeden, Fred L. Troise, and David Keith Todd, *The Water Encyclopedia* (Chelsea, Mich.: Lewis, 1990).

19 Sandra Postel, "Managing Freshwater Supplies," in *State of the World,* ed. Lester R. Brown (New York: Norton, 1985).

20 P. Eaton, "How Bad Is New York's Environment?" *New York Magazine,* Apr. 16, 1990.

21 William J. Flipse et al., "Sources of Nitrate in Groundwater in a Sewered Housing Development, Central Long Island, New York," *Ground Water* 22, no. 4 (1984): 418–26.

22 U.S. Environmental Protection Agency, *National Pesticide Survey: Summary Results of EPA's Na-*

tional Survey of Pesticides in Drinking Water Wells (Washington, D.C.: Environmental Protection Agency, Office of Water, Office of Pesticides and Toxic Substances, 1990), 3.

23 A. M. Petrovic, "Golf Course Management and Nitrates in Groundwater," Golf Course Management, September 1989, 54–65.

24 Arthur J. Gold et al., "Nitrate-Nitrogen Losses to Groundwater from Rural and Suburban Land Uses," Journal of Soil and Water Conservation 45, no. 2 (1990): 305–10; World Resources Institute, in collaboration with the United Nations Environment Programme and the United Nations Development Programme, "Freshwater," in World Resources, 1992–93, ed. Allen L. Hammond (New York: Oxford University Press, 1992).

25 T. G. Morton, Arthur J. Gold, and W. Michael Sullivan, "Influence of Overwatering and Fertilization on Nitrogen Losses from Home Lawns," Journal of Environmental Quality 17, no. 1 (1988): 124–30; Gold et al., "Nitrate-Nitrogen Losses to Groundwater."

26 R. Eugene Turner and Nancy N. Rabalais, "Changes in Mississippi River Quality This Century," BioScience 41, no. 3 (1991): 140–47.

27 Turner and Rabalais, "Mississippi River Quality," 144.

28 U.S. Environmental Protection Agency, National Pesticide Survey, 1–2.

29 Robert F. Carsel and Charles Nicholas Smith, "Impact of Pesticides on Ground Water Contamination," in Silent Spring Revisited, ed. Gino J. Marco, Robert M. Hollingworth, and William Durham (Washington, D.C.: American Chemical Society, 1987).

30 J. M. Halstead, W. R. Kearns, and P. D. Relf, "Lawn and Garden Chemicals and the Potential for Groundwater Contamination," in Proceedings of Ground Water: Issues and Solutions in the Potomac River Basin/Chesapeake Bay Region (Washington, D.C.: National Water Well Association, 1989), 355–69.

31 U.S. Congress, Office of Technology Assessment, Facing America's Trash: What Next for Municipal Solid Waste? OTA-0-424 (Washington, D.C.: U.S. Government Printing Office, 1989); B. Gavitt, "Recycling Clippings Eases Pressure on Landfills," Turf 4, no. 2 (1991): 20–22.

32 U.S. Congress, Office of Technology Assessment, Facing America's Trash.

33 U.S. Congress, Office of Technology Assessment, Facing America's Trash.

34 Paul Ehrlich and Edward O. Wilson, "Biodiversity Studies: Science and Policy," Science 253 (1991): 760.

35 D. L. Cauley, "Urban Habitat Requirements of Four Wildlife Species," in Wildlife in an Urbanizing Environment, ed. John H. Noyes and Donald R. Progulske (Amherst: University of Massachusetts Cooperative Extension, 1974).

36 Kenneth V. Rosenberg, Scott B. Terrill, and Gary H. Rosenberg, "Value of Suburban Habitats to Desert Riparian Birds," Wilson Bulletin 99, no. 4 (1987): 642–54.

37 John T. Emlen, "An Urban Bird Community in Tucson, Arizona: Derivation, Structure, Regulation," Condor 76, no. 2 (1974): 184–97.

38 S. W. Aldrich and R. W. Coffin, "Breeding Bird Populations from Forest to Suburbia after Thirty-seven Years," *American Birds* 34, no. 1 (1980): 3–7; J. H. Falk, "Energetics of a Suburban Lawn Ecosystem," *Ecology* 57 (1976): 141–50.

39 D. N. Jones, "Temporal Changes in the Suburban Avifauna of an Inland City," *Australian Wildlife Research* 8 (1981): 109–19; P. Mason, "The Impact of Urban Development on Bird Communities of Three Victorian Towns—Lilydale, Coldstream and Mt. Evelyn," *Corella* 9, no. 1 (1985): 14–21.

40 Robert C. Tweit and Joan C. Tweit, "Urban Development Effects on the Abundance of Some Common Resident Birds of the Tucson Area of Arizona," *American Birds* 40, no. 3 (1986): 431–36.

41 Emlen, "Urban Bird Community," 184–97.

CHAPTER 5: A NEW AMERICAN LAWN

1   Aurora City Ordinances, Aurora, Colo., 1992.

2   M. Amato, "Novato's Disappearing Lawns: Restating the Case for Lawnless Xeriscape," *Garbage* 2, no. 4 (1990): 34.

3   Texas Water Development Board, *A Homeowner's Guide to Water Use and Water Conservation* (Austin: Texas Water Development Board, 1990).

4   Warren Schultz, *The Chemical-Free Lawn* (Emmaus, Pa.: Rodale, 1989).

5   "A New Look for Lawns," *American Horticulturist* 69, no. 11 (1990): 3.

6   Jay Burnett, "Organic Lawn Care," *Organic Gardening* 37, no. 5 (1990): 70.

7   G. Bugbee, Connecticut Agricultural Experiment Station, interview conducted by J. Greenfeld, February 1991.

8   "How Much Water Does Your Lawn Really Need?" *Sunset,* June 1987, 213–19.

9   Reza Aurasteh, M. Jafari, and L. S. Willardson, "Residential Lawn Irrigation Management (Homeowner Water Management, Utah)," *Transactions of the American Society of Agricultural Engineers* 27, no. 2 (1984): 470–72.

10 Bruce K. Ferguson, "Water Conservation Methods in Urban Landscape Irrigation: An Exploratory Overview," *Water Resources Bulletin* 23, no. 1 (1987): 147–52.

11 U.S. Department of Commerce, *United States Industrial Outlook '90* (Washington, D.C.: U.S. Government Printing Office, 1990).

12 Professional Lawn Care Association of America, News Release, Marietta, Ga., Dec. 1, 1990, 4.

13 Barbara J. Barton, *Gardening by Mail: A Source Book,* 3d ed. (Boston: Houghton Mifflin, 1990); Henry W. Art, *The Wildflower Gardener's Guide,* Midwest, Great Plains, and Canadian Prairies Edition. A Garden Way Publishing Book (Pownal, Vt.: Storey Communications, 1990–91). Five separate editions exist for the Midwest, Great Plains, and Canadian prairies; Pacific Northwest, Rocky Mountain, and West; California, desert Southwest, and northern Mexico; Northeast, mid-Atlantic, Great Lakes, and eastern Canada; and Southeast and Gulf

Coast. Also see Henry W. Art, *The Wildflower Gardener's Guide: 101 Native Species and How to Grow Them* (Pownal, Vt.: Storey Communications, 1990).

CHAPTER 6: THE LAWN AND SUSTAINABILITY

1 The Milford Community Freedom Lawn Initiative is a joint project of the Environmental Concerns Coalition, the Milford Conservation Commission, and the Connecticut Audubon Coastal Center. The coined term *Freedom Lawn* was borrowed from the first edition of this book.

2 Environmental Concerns Coalition, Milford, Conn., *Freedom Lawn* (1998, 1999).

3 Membership information, Garden Club of America, National Headquarters, New York City, November 1999.

4 David J. Eagan and David W. Orr, eds., *The Campus and Environmental Responsibility* (San Francisco: Jossey-Bass, 1992).

5 The complete text of the Talloires Declaration is available at the Association of University Leaders for a Sustainable Future website, <www.ulsf.org/about/tallo.html>.

6 For the complete statement of the Connecticut College Environmental Policy, see <www.camel.conncoll.edu/ccrec/greennet/Environmental_Policy/environmental_policy.html>.

7 Water from the Colorado River is being drawn and stored in reservoirs. However, the water has not been used yet for the city's daily water consumption supply.

8 Lester R. Brown, "Crossing the Threshold: Early Signs of an Environmental Awakening," *WorldWatch,* March–April 1999, 18.

9 Personal communication with Tim Sanderson, November 1998.

10 State of Connecticut, Department of Environmental Protection, *New Corporate Landscape Reduced Toxics: A Pollution Prevention Case Study,* 1996.

11 Steve Grant, "Lawn Gone, But Keep the Faith: Greener Pastures Will Return with Some Rain," *Hartford Courant,* Sept. 16, 1995.

12 State of Connecticut, Department of Environmental Protection, *New Corporate Landscape Reduced Toxics.*

13 Grant, "Lawn Gone, But Keep the Faith."

14 J. Rundquist, "Turning Heads in Raritan: Riot of Wildflowers Delights Workers, Passers-by at Ortho," *Star Ledger,* June 15, 1995.

15 Wildlife Habitat Council, *WHC, Wildlife, and Business/WHC Certification: Program for Corporate Wildlife Habitat,* Silver Spring, Md., 1997.

16 Stevie Daniels, *The Wild Lawn Handbook: Alternatives to the Traditional Front Lawn* (New York: Macmillan, 1995). Also Dr. Darryl Morrison, University of Georgia at Athens, personal communication, October–November 1999.

17 Sandra Postel, "The Politics of Water," *WorldWatch,* July–August 1993.

18 U.S. Department of Commerce, U.S. Census Bureau, *Statistical Abstract of the United States: New Privately Owned Housing Units Started—Selected Characteristics: 1970 to 1996*, no. 1185 (Washington, D.C.: U.S. Government Printing Office, 1998), 717.

19 Robert Zimmerman, Jr., executive director of Charles River Watershed Association (CRWA). Zimmerman's essay "Lawns, Land Use, and Limitations" (1999) was written expressly for this book. Permission was granted to integrate the essay with the suburbanization and hydrology section.

20 F. Herbert Bormann and Gene E. Likens, *Patterns and Process in a Forested Ecosystem: Disturbance, Development, and the Steady State Based on the Hubbard Brook Ecosystem Study* (New York: Springer-Verlag, 1979).

21 Thomas Schueler, executive director, the Center for Watershed Protection, Ellicott, Md.

22 Yale University, School of Forestry and Environmental Studies, Center for Coastal and Watershed Systems, *Water Quality in the Quinnipiac River: A Symposium on the Impact of the Non-Point Source Pollution in the Quinnipiac River Watershed*, Nov. 23, 1998.

23 R. J. Waschbusch, W. R. Selbig, and R. T. Bannerman, *Sources of Phosphorus in Storm Water and Street Dirt from Two Urban Residential Basins in Madison, Wisconsin, 1994–1995*, Water-Resources Investigations Report 99-402 (Middleton, Wis.: U.S. Geological Survey, 1999).

24 Chester L. Arnold, Jr., and C. James Gibbons, "Impervious Surface Coverage: The Emergence of a Key Environmental Indicator," *Journal of the American Planning Association* 62, no. 2 (1996): 253.

25 Personal communication with Larry Coffman, associate director, Department of Natural Resources, Prince George's County Government, Md., 1999.

26 M. Hughes, *Wildflower Manual* (Texas Department of Transportation, Austin, 1993), sec. 3, p. 3.

27 North Carolina Department of Transportation Roadside Environmental Unit, *Wildflowers on North Carolina Roadsides* (Raleigh, N.C., 1998).

28 Personal communication with William Johnson, roadside environmental supervisor, North Carolina Department of Transportation, March 1999. North Carolina Department of Transportation, *North Carolina Department of Transportation Wildflower Program* (July 1998).

29 Daniels, *Wild Lawn Handbook*, 10.

30 Carol Kaesuk Yoon, "Altered Corn May Imperil Butterfly, Researchers Say," *New York Times*, May 20, 1999.

31 Nature Conservancy, Species Report Card, 1997. Jessica Bennett Wilkinson, "The State Role in Biodiversity Conservation," *Issues in Science and Technology* 15, no. 3 (1999): 71.

32 J. Ahern and J. Boughton, "Wildflower Meadows as Sustainable Landscapes," in *The Ecological City: Preserving and Restoring Urban Biodiversity*, ed. Rutherford H. Platt, Rowan A. Rowntree, and Pamela C. Muick (Amherst: University of Massachusetts Press, 1994), 174–75.

33 Hughes, *Wildflower Manual*, sec. 3, p. 3.

34 Hughes, *Wildflower Manual*, sec. 3, p. 3.

35 Personal communication with Melody Hughes, floriculturist, Maintenance Division, Texas Department of Transportation, 1999.

36 Ahern and Boughton, "Wildflower Meadows as Sustainable Landscapes," 175.

37 <www.desertusa.com/Thingstodo/du_ttd_bloom.html>; <www.donaldburger. com/ wfindex.htm>. A Dallas–Fort Worth wildflower photo tour can be found at <www.inter-plaza.com/wildflowers/firststop.htm>.

38 Charles G. Sibley and Burt L. Monroe, Jr., *Distribution and Taxonomy of Birds of the World* (New Haven and London: Yale University Press, 1990), xxi; Larry Master, Natural Heritage Central Databases at the Association for Biodiversity Information, Boston, November 1999; Allan C. Fisher, Jr., "Mysteries of Bird Migration," *National Geographic,* August 1979, 165.

39 "Routes of Migration," in U.S. Department of Interior, U.S. Fish and Wildlife Service, *Migration of Birds: Circular 16,* rev. by John L. Zimmerman, Kansas State University (1998), 53–54.

40 Noreen Damude, "Long Distance Travelers," *Texas Parks and Wildlife,* April 1999, 55.

41 Paul R. Ehrlich, David S. Dobkin, and Daryl Wheye, "Shorebird Migration and Conservation," in *The Birder's Handbook: A Field Guide to the Natural History of North American Birds* (New York: Simon and Schuster, 1988), 121; Paul Kerlinger, "Showdown at Delaware Bay," *Natural History,* May 1998, 56.

42 Ehrlich, Dobkin, and Wheye, *Birder's Handbook,* 121.

43 The Cape May Bird Observatory, <www.njaudubon.org>.

44 Eric Stiles and Lawrence J. Niles, "The New Jersey Division of Fish, Game and Wildlife's Endangered and Nongame Species Program," in *The Cape May Stopover Protection Project: A Partnership in Protection* (Cape May, N.J., 1998). Note: Peter Dunne, director of the New Jersey Audubon Society's Cape May Bird Observatory, estimated the $31 million figure.

45 The Stopover Protection Project, which has become a national model, was a two-and-a-half-year funded partnership among four agencies (the New Jersey Division of Fish, Game, and Wildlife's Endangered and Nongame Species Program, the Nature Conservancy, the New Jersey Audubon Society, and the Association of New Jersey Environmental Commissions). The majority of the funding for the project came from a nonprofit organization, Fund for New Jersey, with additional assistance from the National Fish and Wildlife Foundation.

46 Personal communication with Endangered and Nongame Species Program wildlife biologist and project coordinator Eric Stiles, May 1999; *Cape May Stopover Protection Project.*

47 Iver Peterson, "Zoning for Visitors Who Fly In," *New York Times,* Apr. 11, 1999; Association of New Jersey Environmental Commissions, *Sample Ordinances for Protecting Significant Coastal Habitats* (Mendham, N.J., 1998).

48 Karen Williams is owner of the landscaping for wildlife business Flora for Fauna.

49 D. Dillon, *The White House and President's Park: Comprehensive Design Plan Summary* (Washington, D.C.: U.S. Department of the Interior, National Park Service, 1998).

50 White House Council on Environmental Quality, *Greening of the White House: Six-Year Report,* Report DOE-EE-0209 (November 1999).

Index

nature, 34; and lawn care industry, 51; and air and water pollution, 66; and ecosystems, 69; and water resources, 89, 131; and lawn management, 111–13

Rhode Island, 18

Rhode Island Wild Plant Society, 120, *121*

Ringer Corporation, 56

Riverside, Illinois, 24, *24*

Runoff, 100–101, 133, 136, 141

SALT (Smaller American Lawns Today), 124

Sanderson, Tim, 127–29

Schueler, Thomas, 135

Scientific knowledge, 34, 35, 45, 70

SCOPE (Students Conscious of Protecting the Environment), 124–25

Scotts Company, 55

Sears Roebuck & Company, 129

Seattle University, 125–26

Sediments, 81–82

Seeds: and lawn history, 18, 90; and lawn care industry, 26, 51, 54, 55–56; and Freedom Lawn, 47, 103; and Industrial Lawn, 48, 56; and lawn care companies, 64; ecolawn seed mixes, 96; and birds, 147

Senate, U.S., 76

Septic tanks, 83

Sierra Club, 45

Smaller American Lawns Today (SALT), 124

Snake River, 81

Social costs: and lawn care industry, 1, 55; and lawns, 28; and landscape designs, 141

Sod, 26–27, 48, 61–62, *62*, 64, 90

Soil: and ecosystems, 4, 70; formation of, 4, 62; and lawn care industry, 54; soil biota, 55; composition of, 58; and water resources, 58, 133; and sod production, 61–62; and Industrial Lawn, 65; variety in, 67; and fertilizers, 77; and water pollution, 82, 146; and lawn management, 94, 111; and clover, 96; and Freedom Lawn, 104; and university campuses, 124; and hydrologic cycle, 140; and chemicals, 141; and highway rights-of-way, 143. *See also* Nutrient cycling

Solar energy: and ecosystems, 4–5, *5*, 67, *68–69*; and Freedom Lawn, 65, 104; and grass plants, 70; and growth rates, 75; increased use of, 92;

and lawn management, 94, 111; and lawn reduction, 106

Solid waste: grass clippings as, 52, 60, 85, 89, 98; and fertilizers, 56; and soil formation, 62; and environment, 85; and landscape design, 141; and White House, 157

South, 19, 25

Southeast, *114*

Southwest: and climate, 25; and irrigation, 58, 73, 81; and ecosystems, 67; and birds, 87; and lawn management, *107*, 109–10, *110*

Species diversity: and environment, 4, 45, 85–88, *87;* and Freedom Lawn, 46–47, 56, 96, 103–4, 117, 118; and Industrial Lawn, 49, 88, 89; and pesticides, 79; and extinctions, 85–86; and birds, 86–88, 106, 148–56; increase of, 92; and corporations, 128; and impervious surfaces, 136; and highway rights-of-way, 143, 147; and meadows, 145

Sprawl, 140

Sterling, James, 118, *119*

Stewart, Nancy, 28–29, 32, 33, 45, 51

Stewart, Walter, 28–29, 32, 33, 45, 51

Students Conscious of Protecting the Environment (SCOPE), 124–25

Suburban developments: and Industrial Lawn, 1, 47, 51; and bird migrations, 2; and landscape design, 2; and lawns, 3, *8, 9, 10, 11,* 19–25, 89; and mowers, 25; and species diversity, 87, 88; and water resources, 130–42, *134, 137, 138, 139;* and environmental zoning, 136–42

Sunken fence, 14, 15, *15,* 19

Sustainability: and lawn management, 1; and Freedom Lawn, 2, 116–20, *117, 119;* and ecology, 4; and landscape designs, 115–16; and university campuses, 120, 122–26; and corporations, 126–29; and suburban developments, 130–42; and highway rights-of-way, 142–47; and bird migration, 147–56; and United States, 157–58

Talloires Declaration, 123

Technology: and lawn care industry, 26–27; and nature, 34; and environmental thought, 38, 40; and Industrial Lawn, 49, 51, 65; and irrigation, 59; and industrialization, 90

Texas, 145, 148

Texas Department of Transportation, 143, 146
Texas Water Development Board, 96
Thoreau, Henry David, 36, 40
Tropical forests, 4, 66, 86
Turfgrass Council of North Carolina, 59–60, 72
Turf Resource Center, 53

United States: and lawns, 3, 12, 17–19, 25,
    109–10, 111–13; vegetation of, 25, 26; and
    ecosystems, 67; DDT banning in, 80; and
    environmental activists, 115; and water
    resources, 131; and biodiversity, 145; and
    birds, 148–49, 150; and sustainability, 157–58
United States Golf Association, 52
University campuses, 1, 10, 18–19, 19, 120,
    122–26, 126, 127
University of Arizona, 122, 125
University of Nevada at Las Vegas, 124–25
University of Vermont, 122
University of Virginia, 18–19, 19
University of Wisconsin-Madison, 122
Urban areas: and lawns, 3, 15, 25, 52, 111; and
    ecosystems, 5; and Industrial Revolution, 20,
    33; and zoning laws, 23; and nature, 33–34;
    and air pollution, 45; and irrigation, 59; and
    water resources, 100–101, 131

Van Heeswyk, Anita, 154
Vaux, Calvert, 24
Vietnam War, 84
Village common, 18
Virginia, 52
Vogt, William, 38, 40, 66

Walpole, Horace, 15
Washington, George, 19
Water pollution: and environment, 3, 81–85; and
    ecosystems, 5; regional problems with, 66;
    and Industrial Lawn, 116, 140; and Freedom
    Lawn, 118, 140; and university campuses, 122,
    124; and storm water, 136; and landscape
    design, 141; and meadows, 146, 147
Water resources: and hydrologic cycle, 1–2, 4,
    55, 130–42, 133, 134, 142–43; and environment,
    3, 80–82; and ecosystems, 4, 70, 132, 133;
    human alteration of, 4; and lawn care
    industry, 54, 55; and irrigation, 58–60, 81, 82,
89, 101, 131; and soil, 58, 133; and fertilizers,
    77, 83, 89; and pesticides, 79, 89; and water
    supply, 81, 82, 89, 101, 130–31, 136, 139; and
    lawn management, 94, 100–102, 106, 109, 110,
    111; and grass plant selection, 95–96; and
    gravel driveways, 117; and Industrial Lawn,
    122; access to, 130–31; and suburban
    developments, 130–42, 134, 135, 136, 138, 139;
    and impervious surfaces, 135–36, 134, 136;
    and birds, 149; and White House, 157. See also
    Droughts; Groundwater; Irrigation
Weeds, 48, 54, 55, 57, 65, 77, 98–99
West, 59, 67, 81, 124, 131
WHC (Wildlife Habitat Council), 129–30
White House, 2, 157–58
Whitman, Walt, 9
Whyte, William, 25
Wilderness Society, 45
Wildflowers: and landscape design, 30, 31–32,
    33; and Freedom Lawn, 118, 124; and
    corporations, 127, 128, 128, 129; and highway
    rights-of-way, 143–44, 144, 146, 147. See also
    Patch
Wildlife: and meadows, 29, 145; and land
    management, 45, 151–52; and lawn care
    industry, 54; and pesticides, 77; and species
    diversity, 86; and lawn reduction, 106; and
    Freedom Lawn, 118, 156; and Smaller
    American Lawns Today, 124; and university
    campuses, 126; and corporations, 129–30;
    and bayscapes, 142; and landscape designs,
    152–53, 154, 155. See also Birds; Patch
Wildlife Habitat Council (WHC), 129–30
Williams, Karen, 153, 156
Wilson, Edward, 86
Wood, William, 18
World Trade Organization, 67
Worldwatch Institute, 41

Xeriscapes, 110, 110

Yale University, 122

Zen Buddhism, 109
Zimmerman, Robert, Jr., 133
Zoning laws, 23, 136–42